CLASSIC REPRINTS

Studies in the History of Mathematical Logic

edited by Stanisław J. Surma

An exact reproduction of the text
originally published by
the Polish Academy of Sciences in 1973

Advanced Reasoning Forum
P. O. Box 635
Socorro, NM 87801 USA
www.ARFbooks.org

ISBN 978-1-938421-26-6

POLISH ACADEMY OF SCIENCES
INSTITUTE OF PHILOSOPHY AND SOCIOLOGY

STUDIES IN THE HISTORY OF MATHEMATICAL LOGIC

EDITED BY
STANISŁAW J. SURMA

WROCŁAW • WARSZAWA • KRAKÓW • GDAŃSK
ZAKŁAD NARODOWY IMIENIA OSSOLIŃSKICH
WYDAWNICTWO POLSKIEJ AKADEMII NAUK
1973

This publication was worked out by the staff of the Department of Logic of the Jagiellonian University in Cracow. The papers which it contains were read at Conference for History of Logic organized annualy by the Section of Logic, Institute of Philosophy and Sociology of the Polish Academy of Sciences

Okładkę i obwolutę projektował
Edward Kostka

Printed in Poland

Zakład Narodowy imienia Ossolińskich — Wydawnictwo, Wrocław. Oddział w Krakowie 1973. Nakład 430+120 egz. Objętość ark. wyd. 17,30; ark. druk. 18; ark. form. A1 23,94. Papier druk. sat. kl. III 70 g 70×100. Oddano do składania 8 VII 1972. Podpisano do druku 8 III 1973. Druk ukończono w marcu 1973. Zam. 584/72. Cena zł 52.—
Drukarnia Uniwersytetu Jagiellońskiego w Krakowie, ul. Czapskich 4

CONTENTS

PREFACE

The volume contains seventeen sketches in the history of modern mathematical logic worked out by the members of the Department of Logic of the Jagiellonian University in Cracow. They were presented in 1966–1971 at the consecutive national conferences of the Thematic Group for the History of Logic organized in Cracow by the Department of Logic of the Polish Academy od Sciences.

All these sketches are devoted to various aspects of the completeness of logical calculi and other formalized deductive theories described in literature on mathematical logic. According to this all the said volume is divided into three parts.

The first part is concerned with completeness of the propositional calculi. It includes a description of E. Post's doctoral dissertation including, a.i., originating from E. Post, of the completeness theorem for the classical propositional calculus (the first sketch), a description of the most important methods of the proof of this theorem (the second sketch) and a survey of the methods of the proof of completeness for the classical equivalential propositional calculus (the third and fourth sketches). The next two articles are devoted to the intuitionistic propositional calculus and present the results of A. Kolmogorov and V. Glivenko concerned with formalization of the logic included in Brouwer's program (the fifth sketch) and a detailed proof of the well-known Jaśkowski's matrix criterion for the intuitionistic propositional calculus (the sixth sketch). Here belong also two papers concerning the propositional calculi intermediate between the classical propositional calculus and the intuitionistic propositional calculus which contain a method for axiomatization of the purely implicational Gödel's matrices (the seventh sketch) and a survey of the most important results in the investigations into the intermediate calculi (the eighth sketch). The last article of the first part informs us about the investigations into, so called, Ackermann's rigorous implication (the ninth sketch).

The second part of the volume is concerned with the completeness of the classical first order predicate calculus. It is composed of a detailed description, originating from K. Gödel, of the completeness theorem of the classical first order predicate calculus (the tenth sketch) and a survey of the most important methods of proving this theorem (the eleventh sketch). Here also belongs an article discussing the genesis of the method of Lindenbaum algebra together with the application of this method to the proof of Gödel's completeness theorem (the twelfth sketch), a characterization of the old and new methods of algebraization of quantifiers (the thirteenth sketch) and a review of L. Rieger's achievements in the field of logic (the fourteenth sketch).

The third part of the volume is concerned with the history of Cantor's definition of the set (the fifteenth sketch) and with the set-theoretical reduction of the concept of relation (the sixteenth sketch). This part is closed by an article devoted to the survey of various meanings of the concept of completeness of the formalized deductive theories (the seventeenth sketch).

The said sketches are merely contributions to the discussion. They are an attempt to report in a detailed and synthetic way a broad list of scientific publications. Some of these publications are now difficult to understand, i.a., because of their obsolete symbolism or terminology. The other of the source publications are scattered in the scientific periodicals which are not easily available. In several sketches it was necessary to correct some accepted but false opinions. It has also become necessary to evaluate some facts once again. We think that this necessity has been inspired by the contemporary stage of development of mathematical logic. Taking into consideration the above moments we hope that our sketches will, perhaps, become useful in the preparation of full monographs of particular chapters in the history of mathematical logic.

I wish to give my thanks to Professor Ryszard Wójcicki, the head of the Department of Logic of the Polish Academy of Sciences in Warsaw for the suggestion to publish this volume. I should also like to express my gratitude to Professor Tadeusz Kubiński, Dr Janusz Onyszkiewicz and Dr Kazimierz Wiśniewski for their valuable help in the preparation of this volume. Thanks are also due to Mr Michał Przybyło who read the manuscript making many linguistic remarks.

Stanisław J. Surma

Cracow, August 15th 1971

I. CONTRIBUTIONS TO THE HISTORY OF COMPLETENESS OF THE PROPOSITIONAL CALCULI

STANISŁAW J. SURMA

EMIL L. POST'S DOCTORAL DISSERTATION [1]

1. Introduction

Emil Leon Post (1897–1954) — one of the most eminent representatives of mathematical logic of the period which followed *Principia Mathematica* of A. Whitehead and B. Russell — took his doctor's degree at Columbia University upon submitting his thesis, *Introduction to a general theory of elementary propositions*, published as a report in [17] and then in a more extended form in [19].

The dissertation brought the historically first profound and original metamathematical characteristics of the propositional calculus described in *Principia Mathematica*, providing the consistency proof of that calculus, the proof of its completeness in the two-element Boolean algebra known usually as two-valued algebra of logic, the proof of its functional completeness, together with the description of the classes of functions which are now known as Post's classes, as well as the construction of very generally conceived finitely many-valued generalizations of the algebra of logic nowadays often called Post's algebras.

Thus, Post's dissertation is placed among the most fundamental works concerning propositional calculus described in *Principia Mathematica*, as well as the classical propositional calculus in general: virtually, it closes up the first stage of the research on the logic described in *Principia Mathematica*. In the next stage the first order predicate calculus described in *Principia Mathematica* has been submitted to thorough investi-

---

[1] This paper is an extended version of a lecture given by the author on April 25, 1969 to the XVth Conference of the History of Logic organized in Cracow by the Section of Logic, Polish Academy of Sciences, together with the Department of Logic of the Jagiellonian University.

It was published in Polish in Universitas Iagellonica Acta Scientiarum Litterarumque, CCXXXIII, Schedae Logicae, fasc. V, Cracoviae MCMLXX, 65–70.

gation. As we know, this research has brought most significant results, to a certain degree analogous to those contained in Post's dissertation. Thus, in 1928 D. Hilbert demonstrated the consistency of the first order predicate calculus described in *Principia Mathematica*, in 1930 K. Gödel proved the theorem about the satisfiability of the theses of the first order predicate calculus described in *Principia Mathematica* in every semantic model of that calculus, whereas in 1935 A. Church proved the unsolvability of the problem of provability in this predicate calculus.

Apart from [29] Post in his dissertation refers to works by Lewis [12] and Schröder [22]. He was also indirectly influenced by Sheffer [23] and Nicod [15].

2. Post's completeness theorem for the classical propositional calculus

The dissertation contains the historically first proof of the consistency of the propositional calculus described in *Principia Mathematica* and of its completeness in the two-element Boolean algebra. Post has founded his proof upon the procedure of zero-one verification, which may be traced back as far as to Frege (1879) and to Peirce (1885). It is to Post, beside Łukasiewicz and Wittgenstein, that this procedure to a large extent owes its present-day form. Let us note, particularly, that though lacking the required formal definitions, Post in fact resorted to the concept of the so-called perfect (or canonical) conjunctive-disjunctive normal form, making use of several properties of this concept. I have described in detail Post's method of proving the completeness theorem in [26]. Let us note, moreover, that in his proof Post was the first to apply the now widely known extensionality lemma, which states the invariance of the propositional connectives with the equivalence connective having cited at length its inductive proof.

3. Post's theorem about the functional completeness of the classical propositional calculus

Post has shown that every at least three-argument function of two-valued algebra of logic is definable in this algebra by means of (superpositions of) at most two-argument functions. This result was subsequently generalized in 1935 and 1936 to any finitely many-valued algebras of logic (cf. [27], [28] and [25]), and in 1945 to infinitely many-valued algebras of logic (cf. [24]).

In 1920 Post started research on the structure of the classes of functions of a two-valued algebra of logic closed under superposition. The dissertation brings a concise characteristics of these classes. More abundant information on the subject was provided by Post in [18]. His research was crowned with a monograph [20], published by Post twenty years after his dissertation. The classes of functions in question, now known as Post's classes, are in fact subalgebras of an algebra built up of all two--valued functions and of the operation of superposition performed upon these functions. We shall record it in a more detailed way. Let C be the class of all functions in the two-valued algebra of logic. Let us denote by $*$ the operation, which to each subclass of the class C correlates all superpositions of functions included within this subclass. The operation $*$ has i.a. the following properties, for any $M, N \subset C$,

(i) $M \subset M^*$,
(ii) $(M^*)^* \subset M^*$,
(iii) if $M \subset N$, then $M^* \subset N^*$,
(iv) $M^* \subset \bigcup \{N^* \subset C\colon N$ is the finite subclass of a class $M\}$.

Let $M \subset C$. A subclass N of the class M such that $N^* = M$ is called a basis in M, unless the equality $N_1^* = M$ holds for any proper subclass N_1 of the class N. Post has demonstrated that every Post's class contained in C has a finite basis. The power of the smallest basis of a Post's class is called the degree of this class. Post has described all the Post's classes included in C and he has proved that there are 66 Post's classes of the degrees 1, 2 and 3 together (removing trivial Post's classes we shall get 3 classes of degree 0, 6 classes of degree 1, 12 classes of degree 2 and 20 classes of degree 3 which gives 41) and 8 classes of degree μ for any $\mu \leqslant 4$.

The number of functions and thus also the number of Post's classes increases rapidly with the transition from the two-valued algebra of logic to the many valued ones. This makes relevant the question of transferring Post's result onto many-valued algebras of logic. Post's result is transmitted to the three-valued algebra of logic. It has appeared, namely, that each of Post's classes composed of the functions of the three-valued algebra of logic has a finite basis (cf. [9] and [3]). The hypothesis, however, that each of Post's classes composed of the functions of the m-valued algebra of logic, where $m \geqslant 4$, has a finite basis expounded by Jablonski in [10] has proved false, which was demonstrated in [11].

The most profound result in the theory of Post's classes is undoubtedly the well-known Post criterion about functional completeness. According to this criterion a class F of the functions of the two-valued algebra of logic is a Post's class if and only if each of the classes below

$$F \setminus J, \quad F \setminus H, \quad F \setminus T, \quad F \setminus S, \quad F \setminus L$$

is non-empty, where the symbol $\setminus$ denotes the set-theoretical difference operation and where J is the class of the monotonic functions, H is the class of the functions preserving the constant zero, T is the class of the functions preserving the constant one, S is the class of selfdual functions and L is the class of the linear functions (the definitions of these concepts are to be found, e.g., in [2]). Unfortunately, this criterion is not transmittable to the many-valued algebras of logic. The Post criterion was expounded by Post in [20] and was improved by Jablonski in [8]. Let us note here that Gluškov in [2] on p. 238 brings the erroneous information that the criterion was expounded by Post still in [19].

The theory of Post's classes (built up of functions in two or more valued algebras) has been discussed in numerous publications (e.g., cf. the reviewing articles [10] and [7]). It has found now wide applications, i.a., to the problems of synthesis of realy-circuit systems.

4. Post's *m*-valued systems of logic

The dissertation contains the description of the m-valued functionally complete algebra of logic, where $m < \aleph_0$, and its interpretation in terms of the theory of m-dimensional Euclidean space. It can be noted, moreover, that the conception of Post's many-valued logic was influenced by Lewis' works about strict implication. Incidentally, it was from Lewis that Post has taken the term „two-valued algebra". Post fully realized that he was construing, according to his own expression, a „non-Aristotelean logic". The perusal of his dissertation does not allow to conclude, as it has been repeatedly done by a number of authors (among whom was e.g. Zinovev in [30]), that Post was acting solely under the impact of the „senseless" combinatorial analysis disregarding all extra-calculatory intuitions. However, another critical reflection seems to be pertinent. Post in his dissertation failed to observe altogether that the many-valued constructions are applicable to proofs of independence in propositional calculus. It is a common thought that the true relevance of many-valued logics on extra-formal philosophical grounds was fully understood only by J. Łukasiewicz who independently of Post in 1920 discovered and described in [13] the three-valued algebra of logic, N. B., functionally incomplete (the method of functional completing of the three-valued Łukasiewicz algebra was subsequently shown by J. Słupecki in [24]), and who only in 1922 generalized it to n values not excluding the infinite number of values (cf. [14]). Incidentally the philosophical hopes which Łukasiewicz aimed to combine with the three-valued and then with more-valued logic failed altogether. Nowadays it is just in combinatorial analysis and in abstract algebra as well as in technical applications that many-valued logic seeks confirmation of its relevance.

To present it in some more details the functionally complete m-valued Post's algebra of logic is an algebraic system

$$\langle \{t_1, t_2, \ldots, t_m\}, \quad \{t_1, t_2, \ldots, t_\mu\}, \quad \rightharpoondown_m, \quad \bigvee_m \rangle$$

where
(i) elements of the sequence $\{t_1, t_2, \ldots, t_m\}$ constitute the so-called logical values such that if $i > j$, then $t_i < t_j$,
(ii) elements of the sequence $\{t_1, t_2, \ldots, t_\mu\}$ constitute the so-called distinguished logical values, where $1 \leqslant \mu < m$, and
(iii) $\rightharpoondown_m$ and $\bigvee_m$ being, respectively, cyclic negation and disjunction defined thus:

$$\rightharpoondown_m t_i = t_1 \text{ for } i = m,$$

$$\rightharpoondown_m t_i = t_{i+1} \text{ for } i \neq m,$$

$$t_i \bigvee_m t_j = \max(t_i, t_j).$$

Post has also examined the functionally incomplete m-valued algebras of logic. In particular, he has described the purely implicational algebras defining the m-valued implication with μ distinguished values, symbolically, $\rightarrow^\mu_m$, as follows:

$$p \rightarrow^\mu_m q = \begin{cases} t_1 & \text{for } p \leqslant q, \\ q & \text{for } t_\mu \leqslant p > q, \\ t_{i-j+1} & \text{for } t_\mu > p = t_j > q = t_i. \end{cases}$$

Let us record as an example the matrix definitions of all the implications in 5-valued Post algebra of logic.

$\rightarrow^4_5$	t_1,	t_2,	t_3,	t_4,	t_5
t_1	t_1	t_2	t_3	t_4	t_5
t_2	t_1	t_1	t_3	t_4	t_5
t_3	t_1	t_1	t_1	t_4	t_5
t_4	t_1	t_1	t_1	t_1	t_5
t_5	t_1	t_1	t_1	t_1	t_1

$\rightarrow^3_5$	t_1,	t_2,	t_3,	t_4,	t_5
t_1	t_1	t_2	t_3	t_4	t_5
t_2	t_1	t_1	t_3	t_4	t_5
t_3	t_1	t_1	t_1	t_4	t_5
t_4	t_1	t_1	t_1	t_1	t_5
t_5	t_1	t_1	t_1	t_1	t_1

$\rightarrow^2_5$	t_1	t_2	t_3	t_4	t_5
t_1	t_1	t_2	t_3	t_4	t_5
t_2	t_1	t_1	t_3	t_4	t_5
t_3	t_1	t_1	t_1	t_2	t_3
t_4	t_1	t_1	t_1	t_1	t_2
t_5	t_1	t_1	t_1	t_1	t_1

$\rightarrow^1_5$	t_1	t_2	t_3	t_4	t_5
t_1	t_1	t_2	t_3	t_4	t_5
t_2	t_1	t_1	t_2	t_3	t_4
t_3	t_1	t_1	t_1	t_2	t_3
t_4	t_1	t_1	t_1	t_1	t_2
t_5	t_1	t_1	t_1	t_1	t_1

This example permits us to note that the implication $\rightarrow_m^1$ converges with the implication in the m-valued algebra of Łukasiewicz's logic (cf. [14]) and thus that the implicational Łukasiewicz algebras are special cases of Post implicational algebras. Further, it results from the above example that if in the implication $\rightarrow_m^{m-1}$ the number $(m-1)$ is replaced by the number 1, while the matrix defining $\rightarrow_m^{m-1}$ is left unaltered, we shall get the well-known Heyting-Gödel implication, intermediate between two-valued and intuitionistic implications (cf. [5] and [4]).

So far, the axiomatization by means of finite axiomatics comprised the implicational algebras of Łukasiewicz (cf. e.g., [21]). The problem of the finite axiomatization of the remaining implicational Post algebras remains open.

5. Metamathematical problems in the dissertation

In his dissertation Post has introduced a number of important metal mathematical concepts, among them the present-day concepts of completeness in Post's sense and consistency in Post's sense. In a genera- case these are, evidently, concepts which differ from the usual notions of consistency and completeness. A system is consistent in Post's sense if and only if no single propositional variable can be proved in it as a theorem. A system is complete in Post's sense if and only if it is both maximal and consistent in Post's sense. Further, the dissertation draws a clear line of distinction between the formal description of the theory of propositions, i.e., the formalized system of the propositional calculus and the informal description of that calculus, i.e., the description of the two-valued algebra of logic in terms of the theory of sets. Such an interpretation of the theory of propositions found later a supporter in D. Hilbert (cf. [6]) and, more recently, in P. Novikov (cf. [16]).

6. Some weak or obsolete points in the dissertation

Post does not altogether take into consideration the problem of independence. As a result he fails to notice that the axiomatics of *Principia Mathematica* is a dependent one nor does he try to apply the many-valued constructions to independence proofs. As we know, the idea was first conceived by P. Bernays in his *Habilitationschrift* in 1918, published in 1926 in [1]. Neither does Post concern himself with the properties of the well-known Sheffer's function. The fact that this function and its dual function sometimes called Łukasiewicz's function are the only functions forming one-element bases in the two-valued algebra of logic, was not proved until E. Żyliński published his [31] in 1925. Finally, one certainly cannot agree with the passages in which Post identifies the actual construction of the formalized axiomatic definition of the propositional calculus with the schematic description of this construction.

References

[1] Bernays P.: *Axiomatische Untersuchungen des Aussagenkalküls der „Principia Mathematica".*
Mathematische Zeitschrift, 25(1926), 305–320.
[2] Gluškov V. M.: *Sintez cifrovych avtomatov.*
Moskva 1962.
[3] Gnidenko B. M.: *Nachoždenie poriadkov predpolnych klassov v trechznačnoj logike.*
Problemy Kibernetiki, 8(1962), 341.
[4] Gödel K.: *Zum intuitionistischen Aussagenkalkül.*
Ergebnisse eines mathematischen Kolloquiums, 4(1931–1932) 40.
[5] Heyting A.: *Die formalen Regeln der intuitionistischen Logik.*
Sitzungsberichte der Preussischen Akademie der Wissenschaften. Physikalisch-mathematische Klasse, 1930, 42–56.
[6] Hilbert D., Ackermann W.: *Grundzüge der theoretischen Logik.*
Berlin 1928.
[7] Ivas'kiv J. L., Pospielov D. A., Tošič Ž.: *Predstavlenia v mnogoznačnich logikach.*
Kibernetika, 2(1969), 35–47.
[8] Jablonski S. V.: *O superpoziciach funkcij algebry logiki.*
Matematičeskij Sbornik, 30 (1952), 329–348.
[9] Jablonski S. V.: *O funkcionalnoj polnote v trechznačnom isčislenii.*
Doklady Akademii Nauk SSSR, 95 (6)(1954), 1153–1156.
[10] Jablonski S. V.: *Funkcionalnye postroenija v k-značnich logikach.*
Trudy Matematičeskogo Instituta Akademii Nauk SSSR, 51(1958), 5–142.
[11] Janov J. I., Mučnik A. A.: *O suščestvovanii k-značnich zamknutych klassov, ne imejuščich konečnogo bazisa.*
Doklady Akademii Nauk SSSR, 127(1) (1959), 44–46.
[12] Lewis C. I.: *A survey of symbolic logic.*
Berkeley 1918.
[13] Łukasiewicz J.: *O logice trójwartościowej.*
Ruch Filozoficzny, 5(1920), 170–171.
[14] Łukasiewicz J., Tarski A.: *Untersuchungen über den Aussagenkalkül.*
Comptes Rendus des Séances de la Société des Sciences et des Lettres de Varsovie, Classe III, 23(1930), 30–50.
[15] Nicod J.: *A reduction in the number of primitive propositions of logic.*
Proceedings of the Cambridge Philosophical Society, 19(1917), 32–41.
[16] Novikov P. S.: *Elementy matematičeskoj logiki.*
Moskva 1959.
[17] Post E. L.: *Introduction to a general theory of elementary propositions.*
Bulletin of the American Mathematical Society, 26(1920), 437.
[18] Post E. L.: *Determination of all closed systems of truth tables.*
Ibid.
[19] Post E. L.: *Introduction to a general theory of elementary propositions.*
American Journal of Mathematics, 43(1921), 163–185.

[20] Post E. L.: *The two-valued iterative systems of mathematical logic.*
Annals of Mathematics, Studies, nr 5, Princeton University Press, Princeton, New Jersey 1941.

[21] Rose A.: *Formalization du calcul propositionnel implicatif à m-valeurs de Łukasiewicz.*
Comptes Rendus.
Paris, 243 (18)(1956), 1263–1264.

[22] Schröder E.: *Vorlesungen über die Algebra der Logik (exakte Logik).*
Leipzig, vol. 1(1890), vol. 2, part 1 (1891).

[23] Sheffer H. M.: *A set of five independent postulates for Boolean algebras, with applications to logical constants.*
Transactions of the American Mathematical Society, 14(1913), 481–488.

[24] Sierpiński W.: *Sur les fonctions de plusieurs variables.*
Fundamenta Mathematicae, 33(1945), 169–173.

[25] Słupecki J.: *Der volle dreiwertige Aussagenkalkül.*
Comptes Rendus des Séances de la Société des Sciences et des Lettres de Varsovie, Classe III, 29(1936), 9–11.

[26] Surma S. J.: *A historical survey of the significant methods of proving Post's theorem about the completeness of the classical propositional calculus.*
This volume.

[27] Webb D. L.: *Generation of any n-valued logic by one binary operator.*
Proceedings of the National Academy of Sciences of the USA, 21(1935), 252–254.

[28] Webb D. L.: *The algebra of n-valued logic.*
Comptes Rendus des Séances de la Société des Sciences et des Lettres de Varsovie, Classe III, 29(1936), 153–168.

[29] Whitehead A. N., Russell B.: *Principia Mathematica.* Vol. I.
Cambridge, England, 1910.

[30] Zinovev A.: *Filozoficzne problemy logiki wielowartościowej.*
Warszawa 1963.

[31] Żyliński E.: *Some remarks concerning the theory of deduction.*
Fundamenta Mathematicae, 7(1925), 203–209.

STANISŁAW J. SURMA

A HISTORICAL SURVEY OF THE SIGNIFICANT METHODS OF PROVING POST'S THEOREM ABOUT THE COMPLETENESS OF THE CLASSICAL PROPOSITIONAL CALCULUS [1]

One of the most important theorem relating to the classical propositional calculus states that every tautology is provable by means of the axioms and the rules of this calculus. It was first proved by Emil Leon Post in 1921 in his doctoral dissertation [14]. With the course of time a number of new methods for proving this theorem has been published. Some of these methods were applied also to prove the completeness of some fragmentary systems of the classical propositional calculus as, e.g., the implicational or equivalential propositional calculus. Simultaneously, attempts have been made to find methods for proving the completeness of various non-classical systems of propositional calculus such as the intuitionistic system, the many valued systems of J. Łukasiewicz or the modal systems of Lewis.

In the present paper I shall restrict myself to the survey of the significant methods of proof of the completeness theorem for the classical propositional calculus. For convenience sake I shall concern myself throughout with the classical propositional calculus with the implication ($\rightarrow$) and negation ($\neg$), based upon the substitution and detachment rules. In a few cases I use the concept of syntactic (resp. semantic) equivalence of propositional formulas. As it is known, a formula A is within the discussed propositional calculus syntactically

---

[1] This paper is an extended version of a lecture given by the author on April 27, 1968 to the XIVth Conference of the History of Logic organized in Cracow by the Section of Logic, Polish Academy of Sciencen, together with the Department of Logic of the Jagiellonian University.

A summary of this lecture was published in Polish in Ruch Filozoficzny, 27 (1969), 172–178.

(semantically) equivalent to the sequence of formulas $B_1, B_2, \ldots, B_n$, where $n \geqslant 1$, if and only if all the implications:

$$A \to B_1,$$
$$A \to B_2,$$
$$\ldots\ldots$$
$$A \to B_n,$$
$$B_1 \to (B_2 \to (\ldots (B_n \to A) \ldots))$$

are provable (are tautologies) within that calculus.

1. Post's method

The shortened proof published by Post in [14] in 1921 was the historically first proof of the completeness theorem of the classical propositional calculus. The monograph [19] on p. 259 erroneously attributes the priority of this proof to J. Łukasiewicz. The same erroneous information is also contained in [8]. Post's proof is founded upon the theory of conjunctive-disjunctive normal forms. Post's arguments were used in later years by, i.a., D. Hilbert and P. Bernays in their monograph [6] in 1934, and D. Hilbert and W. Ackermann in the second edition of their monograph [5] in 1937. The first edition of [5] in 1928 does not altogether contain the completeness theorem. Among the more recent monographs, Post's method has been described at length, e.g., in the monograph [1].

To prove the completeness theorem by Post's method we shall have recourse to lemmas formulated below.

LEMMA 1. *If A is a tautology and if $p_1, p_2, \ldots, p_n$ are all the variables occurring in A, then A is semantically equivalent to the formula*

$$(p_1 \vee \neg p_1) \wedge (p_2 \vee \neg p_2) \wedge \ldots \wedge (p_n \vee \neg p_n).$$

The proof of this lemma is trivial.

As we know, the disjunction, in which every argument is either a variable or a negation of a variable, is called an elementary disjunction. The conjunction of elementary disjunctions, equivalent to the given formula A and built up of the same variables as A is called the conjunctive-disjunctive normal form of formula A.

LEMMA 2. *If A is an elementary disjunction built up of the variables $p_1, p_2, \ldots, p_n$, then there exists a conjunctive-disjunctive normal form B built up of exactly the same variables $p_1, p_2, \ldots, p_n$ which is syntactically equivalent to the formula $\neg A$.*

The proof of the lemma is founded upon the theorem of double negation, on the generalized de Morgan's theorems as well as upon the rule of extensionality which stipulates that the fact that formula C is syntactically equivalent to formula D entails that formula Ω is syntactically equivalent to formula $\Omega(C/D)$, in which at least some occurrence of the formula C is replaced by the formula D. The proof of the extensionality rule itself is by induction on the construction of formula Ω.

LEMMA 3. *If* $A_1, A_2, \ldots, A_m$ *are conjunctive-disjunctive normal forms and if* $p_1, p_2, \ldots, p_n$ *are all the variables occurring in these formulas, then there exists a conjunctive-disjunctive normal form* B *built up of exactly the same variables* $p_1, p_2, \ldots, p_n$ *which is syntactically equivalent to the formula*

$$A_1 \vee A_2 \vee \ldots \vee A_m .$$

The proof of the lemma is inductive in view of k. The starting step itself, when $k = 2$, is proved by induction on the number of factors in both components of the disjunction.

LEMMA 4. *If* A *is a conjunctive-disjunctive normal form built up of the variables* $p_1, p_2, \ldots, p_n$, *then there exists a conjunctive-disjunctive normal form* B *built up of exactly the same variables* $p_1, p_2, \ldots, p_n$ *which is syntactically equivalent to the formula* $\rightarrow A$.

The proof of the lemma is founded upon the extensionality rule and upon lemmas 2 and 3.

LEMMA 5. *Every formula* A *is syntactically equivalent to its conjunctive-disjunctive normal form.*

The proof of lemma 5 is by induction on the construction of A and is founded upon lemmas 3 and 4.

Here is the proof of the completeness theorem by Post's method. Let A be a tautology. From the definition of formulas in the propositional calculus it results that in A there occurs a finite number of variables. Let $p_1, p_2, \ldots, p_n$ be all such variables. According to lemma 1 it results from this that A is a formula semantically equivalent to formula

$$\text{(i)} \qquad (p_1 \vee \rightarrow p_1) \wedge (p_2 \vee \rightarrow p_2) \wedge \ldots \wedge (p_n \vee \rightarrow p_n) .$$

Since that latter formula is the conjunctive-disjunctive normal form of formula A, then according to lemma 5 formula A is syntactically equivalent to formula (i). But formula (i) is provable, as we know, in the classical propositional calculus. Thence A is also a provable formula in the classical propositional calculus.

2. Tarski's method

As it results from [9], [12] and [2], the new proof of the completeness theorem was expounded by Alfred Tarski in Warsaw in 1925. It seems, however, that it has not been published anywhere. We know only that Tarski proved the completeness of the purely implicational classical propositional calculus. [22] leads us also to surmise that the later method of Wajsberg was very similar to Tarski's method. This will be mentioned when we discuss Wajsberg's method.

3. Łukasiewicz's method

The author of the next, highly ingenious and original method of proof of the completeness theorem was Jan Łukasiewicz. He first presented it at his lectures in mathematical logic at the University of Warsaw, delivered in the Autumn term of the academic year 1928–29. He published it in 1929 in his textbook [9]. Information on the subject is contained also in [12]. In reference to Łukasiewicz's method see also [10] and [4].

Łukasiewicz carried out a syntactic proof of deductive completeness of the classical implicational-negational propositional calculus. According to his proof from the fact of the existence of formulas unprovable in this calculus which can, however, be consistently added to it, it would result that one of them should be the simplest, and this — as it can be shown — is impossible. We shall describe this proof in somewhat more details. Łukasiewicz founded his argument on the following lemmas.

LEMMA 1. *If A is a formula syntactically equivalent to the sequence of formulas $B_1, B_2, \ldots, B_n$ and for any $i \leqslant n$ either B_i is provable in the classical propositional calculus or when added to this calculus leads up to inconsistency, then either A is provable in the classical propositional calculus, or when added to it leads up to inconsistency.*

Proof. Let us assume that A can be added consistently to the classical propositional calculus. According to the first premise of the lemma and according to the definition of syntactic equivalence for any $i \leqslant n$ formula B_i can be then also consistently added to the classical propositional calculus. Thence it results according to the second premise of our lemma that B_i is also provable in the classical propositional calculus. But, ex hypothesi, the implication $B_1 \to \big(B_2 \to \big(\ldots(B_n \to A)\ldots\big)\big)$ is provable in the classical propositional calculus, thence ultimately A is a formula provable in the classical propositional calculus.

LEMMA 2. *For any formula A there exist formulas $B_1, B_2, \ldots, B_n$ and a propositional variable p such that A is syntactically equivalent to formula* $B_1 \to \big(B_2 \to \big(\ldots(B_n \to \neg(p \to p))\ldots\big)\big)$.

It is easy to see that any formula A is syntactically equivalent to formula $\rightharpoondown A \to \rightharpoondown(p \to p)$. Thence results lemma 2.

LEMMA 3. *If* $A_1, A_2, \ldots, A_n$ *are propositional variables or negations of propositional variables, then either the formula*

(i) $$A_1 \to \Big(A_2 \to \Big(\ldots \big(A_n \to \rightharpoondown(p \to p)\big)\ldots\Big)\Big)$$

is provable in the classical propositional calculus or when added to this calculus leads up to inconsistency.

Proof. Let us assume that all antecedents of formula (i) are either variables or negations of variables. If among the antecedents of formula (i) there are inconsistent formulas, then evidently formula (i) is provable in the classical propositional calculus. If, on the contrary, among the antecedents (i) there are no inconsistent formulas, then for the antecedents which are variables we substitute formula $p \to p$, whereas for the antecedents which are negations of variables we substitute formula $\rightharpoondown(p \to p)$. It is obvious that a formula thus obtained after being added to the classical propositional calculus will lead to inconsistency. Thence it results that also formula (i) after added to the classical propositional calculus will lead to inconsistency. This ends the proof of lemma 4.

LEMMA 4. *If for any* $i \leqslant n$ *a formula* A_i *is neither a variable nor a negation of a variable, then there exist formulas* B *and* C *such that the formula*

(i) $$A_1 \to \Big(A_2 \to \Big(\ldots\big(A_n \to \rightharpoondown(p \to p)\big)\ldots\Big)\Big)$$

is syntactically equivalent either to the formula B *or to the formulas* B *and* C, *where* B *as well as* C *are simpler then (i).*

Let us recall that A is simpler than B, if either A is built of a smaller number of symbols than B or A is built of the same number of symbols as B but A has more antecedents than B.

Proof of lemma 4. A formula A_i fulfilling the premise of the lemma has one of the following forms

(i) $\rightharpoondown\rightharpoondown D$, $D \to E$, $\rightharpoondown(D \to E)$.

Let us observe that:

(ii) $\rightharpoondown\rightharpoondown D \to F$ is syntactically equivalent to a simpler formula $D \to F$;

(iii) $(D \to E) \to F$ is syntactically equivalent to simpler formulas $\rightharpoondown D \to F$ and $E \to F$;

(iv) $\rightharpoondown(D \to E) \to F$ is syntactically equivalent to a simpler formula $D \to (\rightharpoondown E \to F)$.

From (i) — (iv) results the lemma.

On the basis of the above lemmas Łukasiewicz has proved that the classical propositional calculus is complete in the sense that every formula unprovable in this calculus, when added to it, leads up to inconsistency. The proof is as follows. Let us assume that A is not provable in the classical propositional calculus and let us indirectly suppose that A can be added consistently to this calculus. According to lemma 2 there exist formulas $A_1, A_2, \ldots, A_n$ and the variable p such that A is syntactically equivalent to the formula

$$A_1 \to \Big(A_2 \to \big(\ldots (A_n \to \neg(p \to p))\ldots\big)\Big). \tag{i}$$

Thence and from lemma 1 it results that (i) is an unprovable formula and that, simultaneously, it does not lead up to inconsistency. Under lemma 3 it results from this that for some $i \leqslant n$ formula A_i is neither a variable nor a negation of a variable. The set Z composed of the unprovable in classical propositional calculus implications of the form (i) whose antecedents are neither variables nor negations of variables is thus non-empty. Since the set of all formulas of the classical propositional calculus is well-ordered by the relation „A is simpler than B", while the set Z is its subset, then Z contains a minimal element. Let us denote it by C. On the other hand, according to lemma 4, formula C is syntactically equivalent to one or two simpler formulas, i.e., it is not the simplest. The constructed inconsistency ends the proof that every formula unprovable in the classical propositional calculus, when added to it, leads up to inconsistency.

The proof of the completeness theorem by Łukasiewicz's method runs as follows. Let A be a tautology and let us assume indirectly that A is not a provable formula in the classical propositional calculus. From this it follows that A after being added to this calculus leads up to inconsistency. Since every theorem of the classical propositional calculus is a tautology and since A is a tautology, then the set of all tautologies would be inconsistent which is impossible.

Let us add, moreover, that Łukasiewicz successfully applied his method once more in 1939 in [11] where he proved the completeness theorem for the purely equivalential propositional calculus.

4. Kalmar's method

A very natural and elegant method of proof of the completeness theorem was next published by the Hungarian logician and mathematician Laslo Kalmar in [7] in 1935. In later years there were many authors who concerned themselves with Kalmar's method. In particular in 1949

L. Henkin in [3] applied it to prove the completeness theorem for any fragment of the propositional calculus comprising the implication. Availing himself of Henkin's results A. Church applied Kalmar's method in his monograph [2] published in 1956. Kalmar's method was presented also at length by G. Asser in [1].

Kalmar founded his argument on the well-known deduction theorem, which stipulates that to prove an implication it is sufficient to deduce from the antecedents of this implication its consequent, applying only the theorems of the classical propositional calculus, the rule of detachment and the rule of substitution for the variables not occurring in the antecedents of this implication. Further, Kalmar applied the following fundamental lemma.

LEMMA. *Let A be a propositional formula and let $p_1, p_2, \ldots, p_n$ be all the variables occurring in it. Let further f be a valuation function which correlates all propositional variables with the elements of the set $\{0, 1\}$. For any $i \leqslant n$ we put*

$$\overline{p_i} = \begin{cases} p_i, & \text{if } f(p_i) = 1, \\ \rightarrow p_i, & \text{if } f(p_i) = 0. \end{cases}$$

Then (i) if the value of the formula A under the valuation f equals to 1, then A is derivable from the formulas $\overline{p_1}, \overline{p_2}, \ldots, \overline{p_n}$ and (ii) if the value of the formula A under the valuation f equals to 0, then $\rightarrow A$ is derivable from the formulas $\overline{p_1}, \overline{p_2}, \ldots, \overline{p_n}$, applying only the theorem of the classical propositional calculus, the rule of detachment and the rule of substitution for the variables not occurring in the antecedents of the formula A, in case it is an implication.

The proof of the lemma is by induction on the construction of the formula A. The lemma is evident when A is a propositional variable. Let then A be a complex formula. Let us assume that $A = \rightarrow A_1$. If the value of A under the valuation f equals to 1, then the value of A_1 under this valuation equals to 0 and according to the inductive hypothesis the formula A is derivable from the formulas $\overline{p_1}, \overline{p_2}, \ldots, \overline{p_n}$. If, on the contrary, the value of A under the valuation f equals to 0, then the value of A_1 under this valuation equals to 1 and according to the inductive hypothesis the formula A_1 is derivable from the formulas $\overline{p_1}, \overline{p_2}, \ldots, \overline{p_n}$. Then, moreover, the formula $\rightarrow\rightarrow A_1 = \rightarrow A$ is derivable from the formulas $\overline{p_1}, \overline{p_2}, \ldots, \overline{p_n}$. Let, at last, $A = A_1 \rightarrow A_2$. Let us assume that the value of A under f equals to 1. Hence it results that under the same valuation f either the value of A_1 equals to 0, or the value of A_2 equals to 1. If the value of A_1 equals to 0, then according to the inductive hypothesis the formula $\to A_1$ is derivable from $\overline{p_1}, \overline{p_2}, \ldots, \overline{p_n}$ and since the formula $\to A_1 \to (A_1 \to A_2)$ is derivable from any set of formulas, then, in this case, we see that the formula $A = A_1 \to A_2$ is derivable from $\overline{p_1}, \overline{p_2}, \ldots, \overline{p_n}$. If, on the contrary, the value of A_2 equals to 1, then according to the inductive hypothesis the formula A_2 is derivable from $\overline{p_1}, \overline{p_2}, \ldots, \overline{p_n}$ and since the formula $A_2 \to (A_1 \to A_2)$ is derivable from any set of formulas, then, in this case, we see that the formula A is derivable from $\overline{p_1}, \overline{p_2}, \ldots, \overline{p_n}$. Let us further assume that the value of A under f equals to 0. Then the value of A_1 equals to 1 whereas the value of A_2 equals to 0. According to the inductive hypothesis hence it results that both the formulas A_1 and $\to A_2$ are derivable from the formulas $\overline{p_1}, \overline{p_2}, \ldots, \overline{p_n}$. But the formula $A_1 \to (\to A_2 \to \to (A_1 \to A_2))$ is derivable from any set of formulas. Hence by virtue of the detachment rule we get that the formula $\to A = \to (A_1 \to A_2)$ is derivable from $\overline{p_1}, \overline{p_2}, \ldots, \overline{p_n}$, which ends the proof of the lemma.

Now we shall prove the completeness of the classical propositional calculus by Kalmar's method. Let A be a tautology and let $p_1, p_2, \ldots, p_n$ be all the variables occurring in A. Hence it follows that for any valuation f of the variables $p_1, p_2, \ldots, p_n$ the value of A equals to 1 and hence and from lemma 1 it follows that the formula A is derivable from the formulas $\overline{p_1}, \overline{p_2}, \ldots, \overline{p_n}$. According to the definition of the formula $\overline{p_i}$ this means that the formula A is derivable from $\overline{p_1}, \overline{p_2}, \ldots, \overline{p_{n-1}}, p_n$ as well as from $\overline{p_1}, \overline{p_2}, \ldots, \overline{p_{n-1}}, \to p_n$. Hence by virtue of the deduction theorem both the formula $p_n \to A$ and the formula $\to p_n \to A$ are derivable from the formulas $\overline{p_1}, \overline{p_2}, \ldots, \overline{p_{n-1}}$. Since the formula $(p_n \to A) \to ((\to p_n \to A) \to A)$ is provable in the classical propositional calculus, then A is derivable from $\overline{p_1}, \overline{p_2}, \ldots, \overline{p_{n-1}}$. Repeating the above reasoning $(n\text{-}1)$ times we shall conclude that the formula A is provable in the classical propositional calculus.

5. Wajsberg's method

One of the well-known methods of proving the completeness theorem — particularly well-known in Poland — is the method prepared by Mordchaj Wajsberg. Wajsberg was the first to prove by his method in 1931 in [23] the completeness of the three-valued Łukasiewicz implicational-negational propositional calculus. Wajsberg's theorem itself, about the complete axiomatization of the three-valued Łukasiewicz propositional calculus, and his theorem about the complete axiomatization of the n-valued Łukasiewicz propositional calculus, where $(n\text{-}1)$ is a prime number, and Lindenbaum's theorem about the complete axiomatization of the n-valued Łukasiewicz propositional calculus, where n is an arbitrary natural number, were expounded without proof as early as 1926 in [12]. Wajsberg again applied his method in 1936 in [24] to prove the completeness of the purely implicational classical propositional calculus. W. V. Quine in [16] in 1938 extended Wajsberg's method to the propositional calculus with the functors of implication and falsum. The completeness proof by Wajsberg's method for the classical propositional calculus with all the functors was discussed by Schröter in [21] in 1943. Wajsberg's method was also throughly described by Asser in his monograph [1]. In Polish Wajsberg's method was presented, i.a., by W. Sadowski in [20] without mentioning the name of Wajsberg himself.

Wajsberg's method is an inductive one. At the starting step it is proved that the tautologies built of exclusively one propositional variable are provable in the classical propositional calculus. Then from the inductive hypothesis that tautologies built up of n propositional variables are provable in the classical propositional calculus it is deduced that tautologies built up of $(n+1)$ propositional variables are also provable in this calculus. Let us describe Wajsberg's method in detail. We use the symbols $A(p/B)$ to denote the formula obtained from the formula A by substituting the formula B for the variable p. We use also the abbreviation $A \leftrightarrow B$ to denote the formula $\neg((A \to B) \to \neg(B \to A))$.

LEMMA 1. *Formulas of the form*

$$(p \leftrightarrow q) \to (A \leftrightarrow A(p/q)) \quad \textit{and} \quad (p \to \neg q) \to (A \leftrightarrow A(p/\neg q)),$$

are provable in the classical propositional calculus.

The proof of the lemma is inductive in view of the number of all the symbols occurring in the formula A.

LEMMA 2. *Formula of the form*

$$A(p/q) \to (A(p/\neg q) \to A)$$

is provable in the classical propositional calculus.

The proof is founded upon lemma 1.

Let us put now

$$\mathfrak{A} = \{q \to q, \quad \neg(q \to q)\}$$

LEMMA 3. *For any* $A \in \mathfrak{A}$ *there exists* $B \in \mathfrak{A}$ *such that the formula* $\neg A \leftrightarrow B$ *is provable in the classical propositional calculus.*

LEMMA 4. *For any* A, $B \in \mathfrak{A}$ *there exists* $C \in \mathfrak{A}$ *such that the formula* $(A \to B) \leftrightarrow C$ *is provable in the classical propositional calculus.*

The proofs of the lemmas 3 and 4 are founded upon simple theorems of the classical propositional calculus.

LEMMA 5. *For any formula* A *built up exclusively of the variable* p *there exist* $B, C \in \mathfrak{A}$ *such that the formulas*

$$A(p/q \to q) \leftrightarrow B \quad \textit{and} \quad A(p/\neg(q \to q)) \leftrightarrow C$$

are provable in the classical propositional calculus.

The proof of the lemma is inductive in view of the number of all the symbols occurring in the formula A and founded upon the lemmas 3 and 4.

It should be noted that lemma 5, so important for Wajsberg's method had been previously published by Tarski in [22] in 1935. Tarski's lemma, however, was restricted to a system of a purely implicational propositional calculus.

LEMMA 6. *If* A *is a tautology built up exclusively of one propositional variable then* A *is provable in the classical propositional calculus.*

Proof. Let us assume that a tautology A is built up exclusively of one propositional variable p. According to lemma 5 there exist $B, C \in \mathfrak{A}$ such that the formulas

(i) $A(p/q \to q) \leftrightarrow B$ and $A(p/\neg(q \to q)) \leftrightarrow C$

are provable in the classical propositional calculus. Moreover, the formulas (i) are tautologies. Hence it follows that

(ii) $B = q \to q$

Because if it was $B = \neg(q \to q)$, then the formula $A(p/q \to q)$ and together with it also the formula A could not be tautologies, contrary to the assumption. By analogous reasoning we get that

(iii) $C = q \to q$

From (ii) and (iii) and from the fact that the formulas (i) are provable in the classical propositional calculus it results that the formulas

(iv) $A(p/q \to q)$, $A(p/\neg(q \to q))$

are also provable in the classical propositional calculus. According to lemma 2 the formula

(v) $A(p/q \to q) \to (A(p/\neg(q \to q)) \to A)$

is also provable in the classical propositional calculus. From (iv) and (v) it ultimately follows that the formula A is provable in the classical propositional calculus.

LEMMA 7. *If tautologies built up of n propositional variables are provable in the classical propositional calculus, then tautologies built up of $(n+1)$ propositional variables are also provable in this calculus.*

The proof of this lemma is trivial.

6. Łoś's method

H. Rasiowa and R. Sikorski suggested in [18] in 1950 that the proof of the completeness theorem for the classical propositional calculus could be carried out in terms of the theory of ideals (resp. filters) in Boolean algebras. This method was worked out by Jerzy Łoś in [8] in 1951. His method was described in Polish by H. Rasiowa in [17] in 1968.

Łoś's method is founded on the following two lemmas.

LEMMA 1. (*Lindenbaum's lemma on syntactically complete systems*). *If A is a formula unprovable in the classical propositional calculus, then there exists at least one set J including all the formulas provable in the classical propositional calculus, closed under detachment rule, not including the formula A and such that for any formula B either B or $\to B$ belongs to J.*

Proof. For any set of propositional formulas X and for any propositional formula B we put

$$X(B) = \{C: (\to B \to C) \in X\} .$$

Let A be a formula unprovable in the classical propositional calculus. Let us consider a sequence

(i) $A_1, A_2, A_3, \ldots$

consisting of all propositional formulas, and let us define an increasing sequence of sets

(ii) $X_0 = \{C: \to A \to C \text{ is provable in the classical propositional calculus}\}$,

(iii) $X_{n+1} = X_n, \quad \text{if} \quad A_{n+1} \in X_n,$

(iv) $X_{n+1} = X_n(A_{n+1}), \quad \text{if} \quad A_{n+1} \notin X_n.$

Let us put

(v) $J = \bigcup_{i=1}^{\infty} X_i.$

It is easy to verify that the set J has the properties required by lemma 1.

LEMMA 2. *If h is a homomorphic mapping of Lindenbaum algebra of the discussed propositional calculus, i.e., of Boolean algebra built up of the* *set of all propositional formulas and of implications and negations as operations in this algebra, into the well-known zero-one matrix, and if* $h(A) = 0$, *then* A *is not a tautology.*

The proof of lemma 2 is immediate.

And here is the proof of the completeness theorem by Łoś's method. Let A be a tautology and let us assume that A is not provable in the classical propositional calculus. There exists then the set of propositional formulas J with properties described in lemma 1. We define valuation function f putting $f(p) = 1$ for $p \in J$, and $f(p) = 0$ for $p \notin J$, where p is an arbitrary propositional variables. This valuation can be extended to homomorphism w_f which attributes value 0 or 1 to an arbitrary propositional formula in a manner known from the so-called zero-one tables. From this it results that w_f is a mapping of Lindenbaum algebra of the classical propositional calculus into the zero-one matrix such that $w_f(A) = 0$. Following lemma 2 we finally conclude that A is not a tautology.

The set J from lemma 1 is a maximal filter in Lindenbaum algebra of the classical propositional calculus (this algebra being reduced by the relation of equivalence of formulas). A method of proof of the completeness theorem, according to which in the set of all propositional formulas in fact the maximal ideal is constructed, containing the discussed unprovable formula, was published by W. Pogorzelski and J. Słupecki in [13] in 1962. The later method may be additionally characterize by the fact that instead of provability it has recourse to the concept of rejection which we owe to Łukasiewicz, and that it is described in Tarski's axiomatic theory of consequence enriched with axioms for the concept of the logical matrix content.

Let us note further that Asser in [1] 1965 formulated a version of lemma 1, differing from this lemma by the fact that instead of the condition:

for any formula B *either* B *or* $\rightarrow B$ *belongs to* J

occurs in the lemma the following weaker condition:

for any formula B *either* B *belongs to* J, *or* A *is obtainable by means of the rule of detachment from the set* J *supplemented with formula* B.

Having thus weakened lemma 1, we still can prove with its aid the completeness theorem for the classical propositional calculus (cf. Asser [1]). It is also interesting to note that so weakened lemma 1 is applicable to the proof of completeness, for instance, of the finitely many-valued Łukasiewicz propositional calculi (cf., e.g., [15]).

References

[1] ASSER G.: *Einführung in die mathematische Logik.* Teil I.
Leipzig 1959.

[2] CHURCH A.: *Introduction to mathematical logic.* Vol. I.
Princeton 1956.

[3] HENKIN L.: *Fragments of the propositional calculus.*
Journal of Symbolic Logic, 14 (1949), 42–48.

[4] HERMES H., SCHOLZ H.: *Ein neuer Vollständigkeitsbeweis für das reduzierte Fregesche Axiomensystem des Aussagenkalküls.*
Forschungen zur Logik und zur Grundlagen der exakten Wissenschaften, Neue Folge, 1 (1937), 40, Leipzig.

[5] HILBERT D., ACKERMANN W.: *Grundzüge der theoretischen Logik.*
Berlin 1928, second edition, 1937.

[6] HILBERT D., BERNAYS P.: *Grundlagen der Mathematik.* Vol. I.
Berlin 1934.

[7] KALMAR L.: *Über die Axiomatisierbarkeit des Aussagenkalküls.*
Acta Scientiarum Mathematicarum, 7 (1935), 222–243.

[8] ŁOŚ J.: *An algebraic proof of completeness for the two-valued propositional calculus.*
Colloquium Mathematicum, 2 (1951), 236–240.

[9] ŁUKASIEWICZ J.: *Elementy logiki matematycznej.* Skrypt autoryzowany opracował M. Presburger.
Nakładem Komisji Wydawniczej Koła Matematyczno-Fizycznego Słuchaczów Uniwersytetu Warszawskiego, 1929, second edition, Warszawa 1958.

[10] ŁUKASIEWICZ J.: *Ein Vollständigkeitsbeweis des zweitwertigen Aussagenkalküls.*
Comptes Rendus des Séances de la Société des Sciences et des Lettres de Varsovie, Classe III, 24 (1931), 151–183.

[11] ŁUKASIEWICZ J.: *Der Äquivalenzkalkül.*
Collectanea Logica, 1 (1939), 145–169 (Warszawa).

[12] ŁUKASIEWICZ J., TARSKI A.: *Untersuchungen über den Aussagenkalkül.*
Comptes Rendus des Séances de la Société des Sciences et des Lettres de Varsovie, Classe III, 23 (1930), 30–50.

[13] POGORZELSKI W. A., SŁUPECKI J.: *Dowód pełności klasycznego rachunku zdań na gruncie aksjomatycznej metodologii.*
Acta Universitatis Wratislaviensis, 12 (1962), 11–18.

[14] POST E. L.: *Introduction to a general theory of elementary propositions.*
American Journal of Mathematics, 43 (1921), 163–185.

[15] PRUCNAL T.: *Dowód aksjomatyzowalności trójwartościowego implikacyjnego rachunku zdań Łukasiewicza.*
Studia Logica, 20 (1967), 133–144.

[16] QUINE W. V.: *Completeness of the propositional calculus.*
Journal of Symbolic Logic, 3 (1938).

[17] RASIOWA H.: *Wstęp do matematyki współczesnej.*
Warszawa 1968.

[18] RASIOWA H., SIKORSKI R.: *A proof of the completeness theorem of Gödel.*
Fundamenta Mathematicae, 37 (1950), 193–200.

[19] RASIOWA H., SIKORSKI R.: *The mathematics of metamathematics.*
Warszawa 1963, second edition, 1968.

[20] SADOWSKI W.: *Pewien dowód zupełności dwuwartościowego rachunku zdań.*
Studia Logica, 11 (1961), 49–53.

[21] SCHRÖTER K.: *Axiomatisierung der Fregeschen Aussagenkalküle.*
Forschungen zur Logik und zur Grundlagen der exakten Wissenschaften, Neue Folge, 8 (1943), Leipzig.

[22] TARSKI A.: *Über die Erweiterungen der unvollständigen Systeme des Aussagenkalküls.*
Ergebnisse eines mathematischen Kolloquiums, 7 (1934–1935), 51–57.

[23] WAJSBERG M.: *Aksjomatyzacja trójwartościowego rachunku zdań.*
Comptes Rendus des Séances de la Société des Sciences et des Lettres de Varsovie, Classe III, 24 (1931).

[24] WAJSBERG M.: *Metalogische Beiträge.*
Wiadomości Matematyczne, 43 (1937), 1–38.

STANISŁAW J. SURMA

A SURVEY OF THE RESULTS AND METHODS OF INVESTIGATIONS OF THE EQUIVALENTIAL PROPOSITIONAL CALCULUS [1]

1. Introduction

This paper deals with the most important results and methods of investigations of the equivalential propositional calculus and some of its extensions. It contains, a.i., the description of the well-known criteria discovered by S. Leśniewski, M. H. Stone and E. Gh. Mihailescu as well as the achievements of the J. Łukasiewicz's seminar in mathematical logic in the thirties. The last section contains a brief information about the author's deduction theorems valid in the equivalential propositional calculus.

By the letters $p, q, r, \ldots$ we shall denote propositional variables. We shall denote the connectives of equivalence and negation by E and N. Any formulas built up of the variables and connectives will be denoted by $x, y, z, \ldots$. We shall use parenthesis symbolism of Jan Łukasiewicz, presenting by means of Exy and Nx, correspondingly, an equivalence and negation formed from the formulas x and y (cf. [5]).

2. Leśniewski's criterion

The earliest remark about the equivalential theorems of the classical propositional calculus may be found in the first volume of *Principia Mathematica* of A. Whitehead and B. Russell from 1910 (cf. [21]). These authors define the equivalence Epq by means of the formula $KCpqCqp$, where K and C are connectives of conjunction and implication (defi-

[1] This is an extension of a talk presented by the author on April 26, 1970 to the XVIth Conference of the History of Logic organized in Cracow by the Section of Logic, Polish Academy of Sciences, together with the Department of Logic of the Jagiellonian University.

nition *4.01 in [21]), and they mention the theorems *Epp* and *EEpqEqp* stating reflexivity and commutativity of the equivalence (theorems *4.2 and *4.21 in [21]).

The next logician who was interested in this subject was Jan Łukasiewicz. He deductively proved still before 1922 the theorem *EEpEqrEEpqr* stating associativity of the equivalence. This theorem is quoted by A. Tarski in his doctoral dissertation dated 1923 (cf. [18], p. 199).

In 1929 Stanisław Leśniewski published his paper [3]. It is the most important paper which has been written about the equivalential fragment in the classical propositional calculus. Many authors referred to it in the later years. Leśniewski's interests in the equivalential propositional calculus were conditioned by his opinion about the definition as a theorem of the formalized system itself. Leśniewski expressed the definition both in the propositional calculus and in his protothetics in an equivalential form.

In this review we shall discuss Leśniewski's paper [3] in a more detailed way.

In the paper [3] Leśniewski accepts two following axioms

(A1) *EEEprEqpErq*

(A2) *EEpEqrEEpqr*

and he proves that these axioms form a complete system, i.e., from these axioms all the purely equivalential theorems of the classical propositional calculus can be deduced by means of the rule of detachment defined by the scheme

$$\frac{x,\ Exy}{y}$$

where x and y are any equivalential formulas, and by means of the rule of substitution any equivalential formulas for the propositional variables.

To prove this fact Leśniewski forms the so-called now Leśniewski's criterion of decidability according to which any equivalential propositional formula can be deduced from the axioma (A1) and (A2) if and only if every propositional variable occurs in it an even number of times. For proving his criterion Leśniewski deduced from the axioms (A1) and (A2) seventy nine theorems of which he uses, for proving his criterion, only the theorems of numbers 7, 19, 20, 21, 69, 70 and 79. In [4] Leśniewski informs us that in 1929 Jan Łukasiewicz simplified the proofs of the theorems 7, 19, 20, 21, 69, 70 and 79 and deduced all of them in forty eight steps. I quote below a formalized proof of these theorems due to Łukasiewicz.

(Ł1) *EEEqrErqErr*
A1 [p/q, q/r].

In the above notation (Ł1) denotes the number of the step in which the theorem *EEEqrErqErr* was achieved in Łukasiewicz's proof whereas the notation A1 [p/q, q/r] denotes that the above theorem was achieved from the axiom (A1) by a simultaneous substitution of the variable q for the variable p and the variable r for the variable q.

(Ł2) *EEErEqrEErqrEEqrErq*
A1 [p/r, q/Erq, r/Eqr]

(Ł3) *EErEqrEErqr*
A2 [p/r]

(Ł4) *EEqrErq*
Ł2, Ł3

This last notation denotes that the theorem (Ł4) was achieved by applying the rule of detachment to the theorems (Ł2) and (Ł3).

(Ł5) *EEEEprEqrErqEErqEEprEqp*
Ł4 [$q/EEprEqp$, r/Erq]

(Ł6) *Err*
Ł1, Ł4

(Ł7) *EEqrEqr*
Ł6 [r/Eqr]

(Ł8) *EErrErr*
Ł7 [q/r]

(Ł9) *EErqEEprEqp*
Ł5, A1

(Ł10) *EEEqrErqEEpEqrEErqp*
Ł9 [q/Erq, r/Eqr]

(Ł11) *EEpEqrEErqp*
Ł10, Ł4

(Ł12) *EEEpEqrEEpqrEErEpqEpEqr*
Ł11 [$p/EpEqr$, q/Epq]

(Ł13) $EEEEprEqpErqEEqrEEprEqp$
Ł11$[p/EEprEqp, q/r, r/q]$

(Ł14) $EErEpqEpEqr$
Ł12, A2

(Ł15) $EEEqrEEprEqpEEprEEqpEqr$
Ł14$[p/Epr, q/Eqp, r/Eqr]$

(Ł16) $EEEprEEqpEqrEEqpEEqrEpr$
Ł14$[p/Epq, q/Eqr, r/Epr]$

(Ł17) $EEsEEpqrEEpqErs$
Ł14$[p/Epq, q/r, r/s]$

(Ł18) $EEqrEEprEqp$
Ł13, A1

(Ł19) $EEEpEqrEEpqrEEsEEpqrEEpEqrs$
Ł18$[p/s, q/EpEqr, r/EEpqr]$

(Ł20) $EEprEEqpEqr$
Ł15, Ł18

(Ł21) $EEsEprEEqsEqEpr$
Ł20$[p/s, r/Epr]$

(Ł22) $EEEEqrsEErqsEEpEEqrsEpEErqs$
Ł20$[p/EEqrs, q/p, r/EErqs]$

(Ł23) $EEEpEqrEEqprEEsEpEqrEsEEqpr$
Ł20$[p/EpEqr, q/s, r/EEqpr]$

(Ł24) $EEEqrEEprEqpEEsEqrEsEEprEqp$
Ł20$[p/Eqr, q/s, r/EEprEqp]$

(Ł25) $EEEsEpEqrEsEEqprEEtEsEpEqrEtEsEEqpr$
Ł20$[p/EsEpEqr, q/t, r/EsEEqpr]$

(Ł26) $EEEEpEqrsEEpqErsEEtEEpEqrsEtEEpqErs$
Ł20 $[p/EEpEqrs, q/t, r/EEpqErs]$

(Ł27) $EEqpEEqrEpr$
Ł16, Ł20

(Ł28) $EEEqrErqEEEqrsEErqs$
Ł27 $[p/Erq, q/Eqr, r/s]$

(Ł29) $EEEsEEpqrEEpEqrsEEEsEEpqrEEpqErsEEEpEqrsEEpqErs$
Ł27 $[p/EEpEqrs, q/EsEEpqr, r/EEpqErs]$

(Ł30) *EEEqrsEErqs*
Ł28, Ł4

(Ł31) *EEpEEqrsEpEErqs*
Ł22, Ł30

(Ł32) *EEEpEqrEEpqrEEpEqrEEqpr*
Ł31 [*p*/*EpEqr*, *q*/*p*, *r*/*q*, *s*/*r*]

(Ł33) *EEEsEprEEqsEEpqrEEsEprEEsqEEpqr*
Ł31 [*p*/*EsEpr*, *r*/*s*, *s*/*EEpqr*]

(Ł34) *EEpEqrEEqpr*
Ł32, A2

(Ł35) *EEsEpEqrEsEEqpr*
Ł23, Ł34

(Ł36) *EEtEsEpEqrEtEsEEqpr*
Ł25, Ł35

(Ł37) *EEEsEprEEqsEqEprEEsEprEEqsEEpqr*
Ł36 [*p*/*q*, *q*/*p*, *s*/*Eqs*, *t*/*EsEpr*]

(Ł38) *EEsEprEEqsEEpqr*
Ł37, Ł21

(Ł39) *EEsEprEEsqEEpqr*
Ł33, Ł38

(Ł40) *EEsEEpqrEEpEqrs*
Ł19, A2

(Ł41) *EEEsEEpqrEEpqErsEEEpEqrsEEpqErs*
Ł29, Ł40

(Ł42) *EEEpEqrsEEpqErs*
Ł41, Ł17

(Ł43) *EEtEEpEqrsEtEEpqErs*
Ł26, Ł42

(Ł44) *EEEsEqrEEtEEprEqpEstEEsEqrEEtEprEEqpEst*
Ł43 [*p*/*t*, *q*/*Epr*, *r*/*Eqp*, *s*/*Est*, *t*/*EsEqr*]

(Ł45) *EEsEqrEsEEprEqp*
Ł24, Ł18

(Ł46) *EEEsEqrEsEEprEqpEEsEqrEEtEEprEqpEst*
Ł45 [*p*/*t*, *q*/*s*, *r*/*EEprEqp*, *s*/*EsEqr*]

(Ł47) $EEsEqrEEtEEprEqpEst$
Ł46, Ł45

(Ł48) $EEsEqrEEtEprEEqpEst$
Ł44, Ł47

Let us notice that in the above proof the numbers 4, 6, 7, 8, 38, 39 and 48 correspond to the numbers 7, 19, 20, 21, 69, 70 and 79 from the original version of Leśniewski's proof presented in [3].

In the proof of Leśniewski's decision criterion the notion of the order of a given equivalence, defined below, plays a very important part.

Let x be an equivalential formula. We say that x is an equivalence of the order n if there exist equivalential formulas y and z such that

(i) $x = Eyz$,

(ii) each of the formulas y and z is built up of n (not necessarily distinct) variables and

(iii) any variable occurs some number of times in the formula y if and only if this variable occurs as many times as in the formula z.

LEMMA 1. *If x is an equivalence built up of n variables, where $n \geqslant 3$, and p is a variable which occurs in x and if for any $k < n$ and for any equivalence y of the order k the equivalence y is a theorem, then there exists an equivalence z of the order n such that*

(i) *z is a theorem*

and there exists z_1 such that

(ii) $z = Exz_1$

where either $z_1 = p$ or there exists z_2 such that $z_1 = Ez_2p$.

Proof. Let us assume that

(1) x is an equivalence built up of n variables

where

(2) $n \geqslant 3$

(3) p is a variable occurring in x

and that

(4) for any $k < n$ and for any equivalence y of the order k the equivalence y is a theorem.

From (1) and (2) it follows that there exist equivalential formulas x_1, x_2 such that

(5) $x = Ex_1x_2$

where there exist $k_1 < n$ and $k_2 < n$ such that

(6) k_1 is the number of variables occurring in x_1

(7) k_2 is the number of variables occurring in x_2

and

(8) $k_1 + k_2 = n$.

On the basis of (1), (5) and (Ł6) it is easy to notice that

(9) if $p = x_2$, then the formula EEx_1pEx_1p fulfills the conditions (i)–(ii) of the proved lemma.

Let us suppose that

(10.1) p occurs in x_2

where

(10.2) $p \neq x_2$.

Let us denote by y the formula constructed of all the occurrences of all variables in x_2, omitting one occurrence of the variable p. Then

(10.3) y is of the order k_2–1

whereas in view of (7)

(10.4) Ex_2Ey_p is of the order k_2

Hence from (4) and from the fact that $k_2 < n$ we get that

(10.5) Ex_2Eyp is a theorem.

Let us observe that in view of (Ł38)

(10.6) $EEx_2ypEEx_1x_2EEyx_1p$ is a theorem.

On the basis of (1), (5) (10.5) and (10.6) it is easy to notice that

(10.7) $ExEEyx_1p$ fulfills the conditions (i)–(ii) of the proved lemma.

Thus

(10) if p occurs in x_2, where $p \neq x_2$, then the formula $ExEEyx_1p$ fulfills the conditions (i)–(ii) of the proved lemma.

On the basis of the steps (1) and (5) and the theorem (Ł4) it is easy to notice that

(11) if $p = x_1$, then the formula $EEpx_2Ex_2p$ fulfills the conditions (i)–(ii) of the proved lemma.

Let us then assume that

(12.1) p occurs in x_1

and

(12.2) $p \neq x_1$.

Let us denote by z the formula built up of all the occurrences of all variables occurring in x_1, omitting one occurrence of the variable p. Then

(12.3) z is of the order k_1–1

whereas in view of (6)

(12.4) Ex_1Ezp is of the order k_1.

Hence, from (4) and from the fact that $k_1 < n$ we get that

(12.5) Ex_1Ezp is a theorem.

Let us observe that in view of the theorem (Ł39)

(12.6) $EEx_1EzpEEx_1x_2EEzx_2p$ is a theorem.

On the basis of (1), (5), (12.5) and (12.7) we conclude that

(12.7) $ExEEzx_2p$ fulfills the conditions (i)–(ii) of the proved lemma.

Thus

(12) if p occurs in x_1, where $p \neq x_1$, then the formula $ExEEzx_2p$ fulfills the conditions (i)–(ii) of the proved lemma.

From (9)–(12) it follows that there exists an equivalence fulfilling the conditions (i)–(ii) of the proved lemma which ends the proof.

LEMMA 2. *For any* $n > 1$ *if* x *is an equivalence of the order* n, *then* x *is a theorem which can be deduced from the axioms* (A1) *and* (A2) *by means of the rules of detachment and substitution.*

Proof. The lemma will be proved by means of the induction on n. Let us observe that if x is an equivalence of the first order, then according to the definition of the order of equivalence there is a variable p such that $x = Epp$, hence it follows, in view of (Ł6), that x is a theorem, while if x is an equivalence of the second order then, in view of the above mentioned definition there are variables p, q such that either $x = EEppEpp$ or $x = EEpqEpq$ or $x = EEpqEqp$, hence it follows, in view of (Ł4), (Ł7) and (Ł8), that x is a theorem. Thus

(I) if $n \leqslant 2$ and x is an equivalence of the order n, then x is a theorem.

Let us assume inductively that

(II) for any $k < n$ and for any equivalential formula y if y is an equivalential formula of the order k, then y is a theorem

and let us suppose that

(1) x is an equivalence of the order n

where

(2) $n \geqslant 3$.

From (1) and (2) it follows, according to the definition of the order of the equivalence, that there exist a variable p and the equivalential formulas

x_1, x_2 such that

(3) $x = Ex_1x_2$

(4) p occurs in x_1

and

(5) x_1, x_2 have exactly n occurrences of precisely the same variables.

We apply lemma 1 to (II), (2), (4) and (5). We get that there exists an equivalence y such that

(6) y is an equivalence of the order n

(7) y is a theorem

and

(8) $y = Ex_1Ey_1p$

for some equivalence y_1. In view of (1), (3), (4) and definition of the order of equivalence it follows that

(9) p occurs in x_2.

We apply lemma 1 to (II), (2), (5) and (9). We achieve that there exists a formula z such that

(10) z is an equivalence of the order n

(11) z is a theorem

and

(12) $z = Ex_2Ez_1p$

for some equivalence z_1.

From (3), (6), (10) and (12) it follows, according to the definition of the order of equivalence, that

(13) EEy_1pEz_1p is an equivalence of the order n.

Hence and from the mentioned definition

(14) Ey_1z_1 is an equivalential formula of the order $< n$.

Hence and from (II)

(15) Ey_1z_1 is a theorem.

Let us observe that in view of the theorem (Ł48)

(16) $EEx_1Ey_1pEEx_2Ez_1pEEy_1z_1Ex_1x_2$ is a theorem.

Applying to (8), (12), (15) and (16) the rule of detachment we get that

(17) Ex_1x_2 is a theorem

and hence and from (3) it follows that

(18) x is a theorem.

Thus

(III) if x is an equivalence of the order n, where $n \geqslant 3$, then x is a theorem.

From (I), (II) and (III) and from the rule of induction follows the proved lemma.

LEMMA 3. *If every variable occurs in* x *an even number of times then* x *is a theorem which can be deduced from axioms* (A1) *and* (A2) *by means of the rules of detachment and substitution.*

Proof. Let us assume that every variable occurs in x an even number of times. Let us denote by n the number of all occurrences of all variables in x. Obviously, n is, as a sum of even numbers, an even number whereas $\frac{n}{2}$ is a natural number. We construct an equivalence y such that precisely the same variables, which occur in x, occur in y, and such that

(1) y is an equivalence of order $\frac{n}{2}$.

From (1) and from lemma 2 it follows that

(2) y is a theorem.

Let us observe that

(3) Eyx is an equivalence of the order n

that is according to lemma 2

(4) Eyx is a theorem.

From (2) and (4) we get according to the rule of detachment that

(5) x is a theorem

which completes the proof.

LEMMA 4. *If* x *is a theorem which can be deduced from the axioms* (A1) *and* (A2) *by means of the detachment and substitution rules, then every variable occurs in* x *an even number of times.*

Proof. It is easy to observe that every variable occurs an even number of times in the axioms and that the rules of detachment and substitution applied to the formulas in which every variable occurs an even number of times lead us to the formulas in which each variable also occurs an even number of times.

From lemmas 3 and 4 the following immediately results.

LEŚNIEWSKI'S CRITERION. *Any equivalence is a theorem which can be deduced from axioms* (A1) *and* (A2) *by means of the rules detachment and substitution if and only if every variable occurs in it an even number of times.*

3. Achievements in Warsaw School

Some further results in the equivalential propositional calculus were achieved in Warsaw School of Logic in the thirties. In 1932 the paper [19] by Mordchaj Wajsberg appeared in which the author quoted without proof the following complete system for the equivalential propositional calculus which he found still before 1930.

(I) *EEEpqrEpEqr*, *EEpqEqp*

(II) *EEpEqrErEqp*, *EEEpppp*

(III) *EEEpEqrEErssEqp*

(IV) *EEEEpqrsEsEpEqr*

Fifteen letters axioms (III) and (IV) are historically the earliest examples of the unique axioms for the equivalential propositional calculus. In the paper [20] written in 1936 Wajsberg modified the system (I) in the following way

(I′) *EEpEqrEEpqr*, *EEpqEqp*

and he proved that out of each of the system (I)–(IV) and (I′) Leśniewski's axioms (A1) and (A2) can be deduced. Clearly the systems (I) and (I′) are mutually equivalent. We shall now show how to derive the axioms (A1) and (A2) from the system (I).

(L1) *EEpqErEpEqr*
I_1, I_1 [p/Epq, q/r, $r/EpEqr$]

(L2) *EsEEpEqrEEEpqrs*
I_1', L1 [$p/EpEqr$, $q/EEpqr$, r/s]

(L3) *EEsEpEqrEEEpqrs*
I_1' [p/s, $q/EpEqr$, $r/EEEpqrs$], L2

(A1) *EEErpEqrEpq*
L1, L3 [p/r, q/p, r/Eqr, s/Epq]

We shall further show how to derive the system (I′) from the system (II).

(M1) *EEqpErEpEqr*
II_1, II_1 [$p/EpEqr$, q/r, r/Eqp]

(M2) $EpEpEEEpppp$
II_2, M1 $[q/EEppp, r/p]$

(M3) $EEEEppppEpp$
II_1 $[q/p, r/EEEpppp]$, M2

(M4) Epp
II_2, M3

(M5) $ErEpEpr$
M1 $[q/p]$, M4

(M6) $EEqqEEqqEpp$
M4, M5 $[p/Eqq, r/Epp]$

(M7) $EEqqEpp$
M4 $[p/q]$, M6

(M8) $EEqpEErEpEqrEEqpErEpEqr$
M1, M1 $[p/ErEpEqr, q/Eqp, r/Eqp]$

(M9) $EEEqpErEpEqrEErEpEqrEqp$
II_1 $[p/Eqp, q/ErEpEqr, r/EEqpErEpEqr]$, M8

(M10) $EErEpEqrEqp$
M1, M9

(M11) $EEEqrEprEEqpEErEpEqrEEqrEpr$
M1 $[q/ErEpEqr, p/Eqp, r/EEqrEpr]$, M10

(M12) $EEErEpEqrEEqrEprEEqpEEqrEpr$
II_1 $[p/EEqrEpr, q/Eqp, r/EErEpEqrEEqrEpr]$, M11

(M13) $EEqpEEqrEpr$
II_1 $[p/r, q/p, r/Eqr]$, M12

(M14) $EEprEEqrEqp$
II_1 $[p/Eqp, q/Eqr, r/Epr]$, M13

(M15) $EEEqqrEEppr$
M7, M13 $[p/Epp, q/Eqq]$

(M16) $EEEEEpqEpqEpqEEqqEpqEEpqEEqqEpq$
II_2 $[p/Epq]$, M13 $[p/Epq, q/EEEpqEpqEpq, r/EEqqEpq]$

(M17) $EEpqEEqqEpq$
M15 $[p/q, q/Epq, r/Epq]$, M16

(M18) $EEEpqEEqqEpqEEpqEqp$
M13 $[r/q]$, M17

(I_2') *EEqpEqp*
M17, M18

(M19) *EEEpqrEEqpr*
M13 [p/Eqp, q/Epq], I_2'

(M20) *EEEEpqrEEqprEEEpqrErEqp*
M14 [$p/ErEqp$, $q/EEpqr$, $r/EEqpr$], I_2' [p/r, q/Eqp]

(M21) *EEEpqrErEqp*
M19, M20

(M22) *EEEpEqrErEqpEEpEqrEEpqr*
M14 [$p/EEpqr$, $q/EpEqr$, $r/ErEqp$], M21

(I_1') *EEpEqrEEpqr*
I_1', M22

From the system (III) the system (I′) can be derived in the following manner:

(N1) *EEEstEtrErs*
III [$p/EEstEtr$, q/Ers, r/s, s/t], III [p/Est, q/t]

(N2) *EqEEErssEqr*
III [$p/EErss$], N1 [r/q, $s/EEErssEqr$, $t/EErss$]

(N3) *EEEqEErssrq*
N1 [$r/EEqEErssr$, s/q, $t/EErss$], N2 [$q/EqEErss$]

(N4) *EErsEsEErss*
III [p/Ers, q/s, $r/EErss$],
N3 [$q/EErsEsEErss$, $r/EErss$]

(N5) *EEErssr*
N1 [$r/EErss$, s/r, t/s], N4

(I_2') *EErsEsr*
III [p/Ers, q/s], N5 [$r/EErsEsr$, $s/EErss$]

(N6) *EErsEEstEtr*
N1, I_2' [$r/EEstEtr$, s/Ers]

(N7) *EErEpEqrEEpEqrEErss*
N5, N6 [$r/EErss$, s/r, $t/EpEqr$]

(N8) *EEEEpEqrEErssEpqEEpqErEpEqr*
N6 [$r/ErEpEqr$, $s/EEpEqrEErss$, t/Epq], N7

(N9) *EEpqErEpEqr*
III, N8

(N10) *EEEpqErEpEqrEErEpEqrEEEpqrr*
N5 [*r*/*Epq*, *s*/*r*], N6 [*r*/*EEEpqrr*, *s*/*Epq*, *t*/*ErEpEqr*]

(N11) *EErEpEqrEEEpqrr*
N9, N10

(N12) *EEEEpqrrErEpEqr*
I_2' [*r*/*ErEpEqr*, *s*/*EEEpqrr*], N11

(I_1') *EEpEqrEEpqr*
N1 [*r*/*EpEqr*, *s*/*EEpqr*, *t*/*r*], N12

From the system (IV) the axioms (A1) and (A2) can be derived in the following manner:

(01) *EEsEpEqrEEpqErs*
IV, IV [*p*/*Epq*, *q*/*r*, *r*/*s*, *s*/*EsEpEqr*]

(02) *EEEpqrEsEsEpEqr*
01, 01 [*p*/*Epq*, *q*/*r*, *r*/*s*, *s*/*EsEpEqr*]

(03) *EEssEEpEqrEEpqr*
01 [*p*/*s*, *q*/*s*, *r*/*EpEqr*, *s*/*EEpqr*], 02

(A2) *EEpEqrEEpqr*
01 [*q*/*p*, *r*/*p*, *s*/*Epp*], 03 [*s*/*EEppEpEpp*]

(04) *EEEpEqrEpqr*
A2, A2 [*p*/*EpEqr*, *q*/*Epq*]

(05) *ErEpEEqrEpq*
A2 [*q*/*Eqr*, *r*/*Epq*, *s*/*r*], 04

(06) *EErpEEqrEpq*
A2 [*p*/*r*, *q*/*p*, *r*/*EEqrEpq*], 05

(A1) *EEErpEqrEpq*
A2 [*p*/*Erp*, *q*/*Eqr*, *r*/*Epq*], 06

In spite of the fact that Wajsberg published his first achievements in the field of the equivalential propositional calculus in 1932 he had been interested in this problem much earlier. This is confirmed by the fact that in a note to the first part of his paper [20] Wajsberg said that he presented the first system of axioms for the equivalential propositional calculus still in 1925 and informed Leśniewski about it in a letter.

Some further information about the equivalential propositional calculus can be obtained from a review article of Bolesław Sobociński [15] written in 1932. The author informs us about a number of the unique axioms for the equivalential propositional calculus which were found in 1930. They are quoted below in the order of being discovered.

(V) $EEsEpEqrEEpqErs$

(VI) $EEpEqrEEqEsrEsp$

(VII) $EEpEqrEEqErsEsp$

(VIII) $EEpEqrEEpErsEsq$

(IX) $EEpEqrEEpEsrEsq$

(X) $EEpEqrEEpErsEqs$

Axioms (V) and (VII) were discovered by Łukasiewicz, axiom (VI) was discovered by J. Bryman, axioms (VIII) — (X) were discovered by Sobociński. In the paper [15] Sobociński says that the problem of the shortest unique axiom for the equivalential propositional calculus was still not solved in the Warsaw School of Logic at this time.

4. Mihailescu's normal forms

In 1937 the paper [8] of the Rumanian logician Eugene Gh. Mihailescu appeared. The author accepts the axiom system (I) in spite of the fact that Wajsberg's papers [19] and [20] seem unknown to him. He considers the axiom system (I) much simpler than Leśniewski's system of axioms. From the axioms (I) he deduces ninety three equivalential theorems of the classical propositional calculus. On that basis he proves by means of induction on n the following two lemmas:

LEMMA 5. *Every equivalential propositional formula containing* $2n$ *of the propositional variables of the shape* p_i, q_i *(where* $i = 1, 2, \ldots, n$*) is deductively equivalent to a normal form*

$$(\mathrm{N}_1) \qquad \underbrace{EE\ldots}_{n\ \text{times}} Ep_1q_1Ep_2q_2\ldots Ep_{n-1}q_{n-1}Ep_nq_n$$

in which the variables p_i, q_i *may be arbitrarily permuted.*

LEMMA 6. *Every equivalential propositional formula containing* $2n+1$ *of the propositional variables of the shape* p_i, q_j *(where* $i = 1, 2, \ldots n$

and $j = 1, 2, \dots n, n+1$) *is deductively equivalent to a normal form*

$$(\mathrm{N}_2) \qquad \underbrace{EE \dots}_{n\ times} Ep_1q_1Ep_2q_2 \dots Ep_{n-1}q_{n-1}Ep_nEq_nq_{n+1}$$

in which the variables p_i, q_j *may be arbitrarily permuted.*

From lemmas 5 and 6 results the following:

LEMMA 7. *If an equivalential formula contains each of the propositional variables an even number of times, then it can be proved from axioms* (I) *by means of the rules of detachment and substitution. If, on the contrary, an equivalential formula contains at least one of the variables an odd number of times, then it leads to a contradiction.*

Lemma 7 shows that the axiom system (I) is deductively complete. Thus is presented a method of the completeness proof worked out by Mihailescu which is based upon reducing any equivalential formula to a normal form (N_1) or (N_2).

Let us still observe that in the paper [8] Mihailescu presents a matrix proof of consistency and independence of the axiom system (I). Leśniewski himself was not interested in this problem.

5. Stone's criterion for the equivalential calculus

In the paper [16] written in 1937 M. H. Stone offered a new decision criterion for the equivalential propositional calculus. Namely he proved the following:

LEMMA 8. *The set of all equivalential formulas with an operation E characterized by the axioms* (A1) *and* (A2) *and with the relation $\equiv$ defined by the condition*

(i) $x \equiv y$ *if and only if Exy is the theorem deducible from the axioms* (A1) *and* (A2) *by means of the rules of detachment and substitution for any equivalential formulas x and y*

forms an additive Abelian group in which every element is of the rank 2.

It has been noted in the literature (cf. e.g., [10]) that Stone's criterion is less convenient than the unusually simple Leśniewski's criterion. However, here it may be well to add that Stone's criterion immediately entails that of Leśniewski's and such a method of proving Leśniewski's criterion is undoubtedly the simplest one.

6. Łukasiewicz's achievements

In the paper [6] written in 1939 and containing the results from 1933 Łukasiewicz quotes eleven letters single axiom of the equivalential propositional calculus:

(XI) $EEpqEErqEpr$

and he proves that this is the shortest axiom as follows. Clearly, every thesis of the equivalential propositional calculus consists on an odd number of letters. Hence all theses which number less than eleven letters must number 1, 3, 5, 7, or 9 letters. On the basis of Leśniewski's decision criterion one can conclude that no thesis of the equivalential propositional calculus may consist of 1, 5, or 9 letters. The only thesis of 3 letters is

(0) Epp

Theses of 7 letters which come under consideration are the following fifteen theses:

(1.1) $EEEppqq$
(1.2) $EEEpqpq$
(1.3) $EEEpqqp$
(2.1) $EEpEpqq$
(2.2) $EEpEqpq$
(2.3) $EEpEqqp$
(3.1) $EEppEqq$
(3.2) $EEpqEpq$
(3.3) $EEpqEqp$
(4.1) $EpEEpqq$
(4.2) $EpEEqpq$
(4.3) $EpEEqqp$
(5.1) $EpEpEqq$
(5.2) $EpEqEpq$
(5.3) $EpEqEqp$

Now Łukasiewicz shows that none of theses (0) — (5.3) axiomatize the equivalential propositional calculus. Namely, the matrix $\langle\{1, 2\}, \{1\}, \underline{E}\rangle$ where $\underline{E}$ is defined by means of the tableau

$\underline{E}$	1	2
1	1	2
2	1	1

satisfies theses (0), (1.1), (3.1), (3.2), (4.1), (4.3), (5.1), (5.2), and (5.3) but it does not satisfy axiom (XI). The matrix $\langle\{1, 2, 3, 4\}, \{1\}, \underline{E}\rangle$ where $\underline{E}$ is defined as follows

$\underline{E}$	1	2	3	4
1	1	2	3	4
2	4	1	1	3
3	2	4	1	1
4	3	1	2	1

satisfies (1.2) and (2.2) but it does not (XI). The matrix $\langle\{1, 2, 3\}, \{1\}, \underline{E}\rangle$ where

$\underline{E}$	1	2	3
1	1	2	3
2	2	1	2
3	3	3	1

satisfies (1.3) but not (XI). The matrix $\langle\{1, 2, 3\}, \{1\}, \underline{E}\rangle$ where

$\underline{E}$	1	2	3
1	1	2	3
2	2	1	3
3	3	2	1

satisfies (2.1) and (2.3) but not (XI). The matrix $\langle\{1, 2, 3\}, \{1\}, \underline{E}\rangle$ where

$\underline{E}$	1	2	3
1	1	2	3
2	2	1	2
3	3	2	1

satisfies (3.3) but not (XI). Finally the matrix $\langle\{1, 2, 3, 4\}, \{1\}, \underline{E}\rangle$ where

$\underline{E}$	1	2	3	4
1	1	4	2	4
2	3	1	4	1
3	3	1	1	2
4	4	3	3	1

satisfies (4.2) but not (XI). From this it follows that none of theses (0) — (5.3) can be the single axiom.

Besides Łukasiewicz describes a new simple method for proving deductive completeness of the equivalential propositional calculus. This method is based upon the fact that every equivalential formula is either a propositional variable or starts with the symbol E and it consists in considering all the cases of equivalential formulas (there are eight of them in Łukasiewicz's [6]) with the discussion whether the said case is deducible from (XI) or when attached to (XI) leads to a contradiction.

On the basis of Leśniewski's decision criterion Łukasiewicz in the paper [6] gives a syntactic proof of the consistency of the equivalential propositional calculus.

On the basis of Leśniewski's decision criterion Łukasiewicz gives in the paper [6] a syntactic proof of the consistency of the equivalential propositional calculus.

Here is the proof that the axiom (XI) comprehends all theses of the equivalential propositional calculus.

(P1) *EEsEErqEprEEpqs*
XI, XI [*p*/*Epq*, *q*/*EErqEpr*, *r*/*s*]

(P2) *EEpqEpq*
XI, P1 [*s*/*Epq*]

(P3) *EErEpqEEpqr*
XI [*p*/*Epq*, *q*/*Epq*], P2

(P4) *EEErqEprEpq*
XI, P3 [*p*/*Erq*, *q*/*Epr*, *r*/*Epq*]

(P5) *Epp*
P2 [*q*/*p*], P4 [*q*/*p*, *r*/*p*]

(P6) *EEpqEqp*
XI [*p*/*q*, *r*/*p*], P5 [*p*/*q*]

(P7) *EErEqpEEpqr*
XI [*p*/*Epq*, *q*/*Eqp*], P6

(P8) *EEEpqrErEqp*
P6 [*p*/*ErEqp*, *q*/*EEpqr*], P7

(P9) *EErqEEprEpq*
P1 [*s*/*EEprEpq*, *r*/*p*, *p*/*r*], P8 [*q*/*r*, *r*/*Epq*]

(P10) *EEpEpqq*
P4 [*r*/*Epq*, *p*/*EpEpq*], P9 [*r*/*Epq*]

(P11) *EqEpEpq*
P6 [*p*/*EpEpq*], P10

(P12) *EErqEEpEpqr*
XI [*p*/*EpEpq*], P10

(A2) *EEpEqrEEpqr*
P1 [*q*/*Eqr*, *r*/*q*, *s*/*EEpqr*], P12 [*p*/*q*, *q*/*r*, *r*/*Epq*]

(P13) *EEEpqrEpEqr*
P6 [*p*/*EpEqr*, *q*/*EEpqr*], A2

(P14) *EEpEqrEpErq*
P8 [*p*/*r*, *r*/*p*], P8 [*p*/*Erq*, *q*/*p*, *r*/*EpEqr*]

(A1) *EEEprEqpErq*
P4 [*p*/*q*, *q*/*r*, *r*/*p*], P14 [*p*/*EEprEqp*]

Besides the axiom (XI) Łukasiewicz has found in [6] two other theses of eleven letters which can be likewise postulated as single axioms of the equivalential propositional calculus. These are

(XII) *EEpqEEprErq*

and

(XIII) *EEpqEErpEqr*

From (XII) the axiom (XI) can be derived in the following manner:

(Q1) *EEEpqsEsEEprErq*
XII, XII [*p*/*Epq*, *q*/*EEprErq*, *r*/*s*]

(Q2) *EEsEEprErqEEEpqrErs*
Q1, Q1 [*p*/*Epq*, *q*/*s*, *s*/*EsEEprErq*]

(Q3) *EEEpqrErEpq*
XII, Q2 [*s*/*Epq*]

(Q4) *EEEprErqEpq*
XII, Q3 [*r*/*EEprErq*]

(Q5) *EpErEpr*
XII [*q*/*Epr*], Q4 [*q*/*ErEpr*, *r*/*Epr*]

(Q6) *EEpqEqErEpr*
XII [*q*/*ErEpr*, *r*/*q*], Q5

(Q7) *EEqErEprEEprErq*
Q1 [*s*/*EqErEpr*], Q6

(Q8) *EEprErp*
Q5, Q7 [*q*/*p*]

(Q9) *EEqpEEprErq*
Q1 [*s*/*Eqp*]. Q8 [*r*/*q*]

(Q10) *EEEpqrErEqp*
Q2 [*s*/*Eqp*], Q9

(XI) *EEpqEErqEpr*
Q4, Q10 [*p*/*Epr*, *q*/*Erq*, *r*/*Epq*]

From (XIII) the axiom (XI) can be derived in the following manner:

(R1) *EEsEpqEEErpEqrs*
XIII, XIII [*p*/*Epq*, *q*/*EErpEqr*, *r*/*s*]

(R2) $EEEsErpEEqrsEpq$
XIII, R1 $[p/Erp, q/Eqr, r/s, s/Epq]$

(R3) $EqErEqr$
R1 $[p/Eqr]$, R2 $[p/q, q/ErEqr, r/Eqr]$

(R4) $EEEsrEEqrsq$
R1 $[p/r, q/Eqr, r/s, s/q]$, R3

(R5) Eqq
R1 $[p/q, r/q]$, R4 $[q/Eqq, r/Eqq]$

(R6) $EErqEqr$
XIII $[p/q]$, R5

(R7) $EEErqEprEpq$
R1 $[p/q, q/p, s/Epq]$, R6 $[r/p]$

(XI) $EEpqEErqEpr$
R6 $[q/Epq, r/EErqEpr]$, R7

However the formulas (XI), (XII) and (XIII) are not the only eleven letters single axioms for the equivalential propositional calculus. As [7] informs us it has been proved by C. A. Meredith in 1951 that the formulas

(XIV) $EEEpqrEqErp$

and

(XV) $ErEEqErpEpq$

are also eleven letters single axioms of the system. We shall show how to derive (XIII) from the axiom (XIV).

(S1) $ErEEqErpEpq$
XIV, XIV $[p/Epq, q/r, r/EqErp]$

(S2) $EEsEEEEpqrEqErptEts$
XIV, S1 $[p/t, q/s, r/S1]$

(S3) $EEErpEEpqrq$
S1 $[p/Erp, q/EEpqr, r/q]$, S2 $[s/q, t/EErpEEpqr]$

(S4) $EEEpqqp$
S1 $[r/EEpqq]$, S3 $[p/q, q/EEEpqqp, r/Epq]$

(S5) $EqEpEpq$
XIV $[p/Epq, r/p]$, S4

(S6) $EErEsEsrEErEsEsrEqEpEpq$
S5, S5 $[p/ErEsEsr, q/\mathrm{S5}]$

(S7) $EErEsEsrEqEpEpq$
S5 $[p/s, q/r]$, S6

(S8) $EEsEsrEEqEpEpqr$
XIV $[p/r, q/EsEsr, r/EqEpEpq]$, S7

(S9) $EEsrEEEqEpEpqrs$
XIV $[p/s, q/Esr, r/EEqEpEpqr]$, S8

(S10) $EEEqEpEpqEEEqEpEpqrsEsr$
S9, S9 $[r/EEEqEpEpqrs, s/Esr]$

(S11) $EEsEtEtsEEsEtEtsEEErpEEpqrq$
S3, S5 $[p/EsEtEts, q/\mathrm{S3}]$

(S12) $EEsEtEtsEEErpEEpqrq$
S5 $[p/t, q/s]$, S11

(S13) $EqEEEpEprqr$
S10 $[q/r, r/EEEpEprqr, s/q]$, S12 $[p/EpEqr, s/r, t/p]$

(S14) $EEsEEqEEEpEprqrtEts$
S1 $[p/t, q/s, r/\mathrm{S12}]$, S12

(S15) $EErqEEpEprq$
S1 $[p/r, r/EEpEprq]$, S14 $[s/EEpEprq, t/Erq]$

(S16) $EEsEsEEpqrEqErp$
XIV, S15 $[p/s, q/EqErp, r/EEpqr]$

(S17) $EEsEEpqrEEqErps$
XIV $[p/s, q/EsEEpqr, r/EqErp]$, S16

(S18) $EEEprEqpErq$
S15, S17 $[q/Epr, r/q, s/Erq]$

(XIII) $EEqpEErqEpr$
XIV $[p/Epr, q/Eqp, r/Erq]$, S18

Let us observe that the axiom (XIV) can be derived from the axiom (XV) in the following manner:

(T1) $EEEqErpEErEEqErpEpqEEpqrEEEpqrEqErp$
XV, XV $[p/EEpqr, q/EqErp]$

(XIV) $EEEpqrEqErp$
XV $[p/Epq, q/r, r/EqErp]$, T1

This proves that (XIV) and (XV) are equivalent each other.

A little later it has been proved by C. A. Meredith that the formulas:

(XVI) *EpEEqErpEqr*

(XVII) *EEpEqrErEpq*

(XVIII) *EEpqErEEqrp*

(XIX) *EEpqErEErqp*

(XX) *EEEpEqrrEqp*

and

(XXI) *EEEpEqrqErp*

are also eleven letters single axioms for equivalential propositional calculus (cf. [7]).

Notice that full description of the set of all eleven letters single axioms for this calculus is not done yet.

7. Rasiowa's calculus

In 1955 there appeared Helena Rasiowa's paper [13] in which the results from 1945 were presented. On the basis of Leśniewski's criterion the author shows that the following modification of the well-known Tarski-Bernays' axiom system for the classical implicational propositional calculus

(XXII) *EEpqEEqrEpr*, *EpEqEqp*, *EEEpqEEpqpp*

axiomatizes the Cartesian product of the implicational matrix

$$\{\{0,1\},\{1\},\underline{C}\}$$

by the equivalential matrix

$$\{\{0,1\},\{1\},\underline{E}\}$$

In Rasiowa's system the connective E may be understood both as the equivalence and as the implication.

8. Mihailescu's criterion for the equivalential-negational calculus

Besides the purely equivalential propositional calculus some investigations into the equivalential calculus with other connectives were made. Attention was paid to both the functionally incomplete and functionally complete enrichments of this calculus. Obviously, much more interesting are the functionally incomplete enrichments of the fragments of the classical equivalential propositional calculus. Here are all the functionally incomplete extensions of the classical equivalential calculus based upon the independent systems of connectives

(i) system with negation N,
(ii) system with non-equivalence E',
(iii) system with disjunction A,
(iv) system with conjunction K,
(v) system with implication C,
(vi) system with counter-implication C'

and

(vii) system with falsum symbol O.

Systems (i) and (ii) were examined in the most detailed way. Some more attention was devoted to the systems (iii) and (iv). But the other systems (iv) — (vii) were not so far taken into consideration.

As far as the equivalential-negational fragment of the classical propositional calculus is concerned we can find already in [21] the theorems *EEpqENpNq* and *EpNNp* (theorems *4.11 and *4.13 in [21]).

As W. V. Quine informs us (cf. [10], [11]) the next logician interested in the equivalential-negational propositional calculus was probably J. C. C. McKinsey. According to Quine still before 1937 McKinsey told him about a generalization of Leśniewski's decision criterion for the equivalential-negational propositional calculus. In accordance with this criterion any equivalential-negational propositional formula is deducible from the axioms of the equivalential-negational propositional calculus if and only if both every variable and the connective of negation N occur in it an even number of times.

However, it is only E. Gh. Mihailescu's paper [9] written in 1937 which fully discusses the properties of the equivalential-negational propositional calculus. Mihailescu accepts the axiom system (I) which he supplements by the axiom

(XXIII) *EENpNqEpq*

Out of this system of axioms he deduces two hundred and twenty equivalential and equivalential-negational theorems. Then he proves by means of the induction on n that the following lemma holds:

LEMMA 9. *Every equivalential-negational propositional formula is equivalent to one of the four normal forms*

$$(\mathrm{N}_1)\quad \underbrace{EE\ldots}_{n\ times} Ep_1q_1Ep_2q_2\ldots Ep_{n-1}q_{n-1}Ep_nq_n$$

(N$_2$) $\underbrace{NEE \ldots E}_{n \text{ times}} p_1 q_1 E p_2 q_2 \ldots E p_{n-1} q_{n-1} E p_n q_n$

(N$_3$) $\underbrace{EE \ldots E}_{n \text{ times}} p_1 q_1 E p_2 q_2 \ldots E p_{n-1} q_{n-1} E p_n E q_n q_{n+1}$

(N$_4$) $\underbrace{NEE \ldots E}_{n \text{ times}} p_1 q_1 E p_2 q_2 \ldots E p_{n-1} q_{n-1} E p_n E q_n q_{n+1}$

depending on the fact that the formula contains:

(i) *the even number of propositional variables and the even number of the connectives N,*

(ii) *the even number of propositional variables and the odd number of the connectives N,*

(iii) *the odd number of propositional variables and the even number of the connectives N or*

(iv) *the odd number of propositional variables and the odd number of the connectives N,*

where the variables p_i and q_j which occur in the formulas (N$_1$) — (N$_4$) *can be arbitrarily permuted.*

On the basis of the lemma 9 Mihailescu proves the following main lemma

LEMMA 10. (i) *If an equivalential-negational formula contains an even number of times each of the propositional variables and an even number the connectives N, then it may be deduced from the axioms* (I) *and* (XXIII),

(ii) *if this formula contains an even number of times each of the propositional variables and an odd number of connectives N, then it cannot be deduced from the axioms* (I) *and* (XXIII) *and at the same time attached to this axiom system does not lead to a contradiction and*

(iii) *if, at least, this formula contains an odd number of propositional variables, then after having attached it to the axiom system* (I) *and* (XXIII) *we get a contradiction.*

Proof. Let x be an equivalential-negational formula containing an even number of variables and an even number of connectives N. Then x is equivalent to a normal form of the type (N$_1$), that is, the formula $E(\mathrm{N}_1)x$ may be deduced from the axioms (I) and (XXIII). But (N$_1$) is deducible from the axioms (I). Therefore x may also be deduced from the axioms (I) and (XXIII). The cases (ii) and (iii) are proved in an analogous way.

From lemma 10 it follows that the equivalential-negational calculus is deductively incomplete but decidable fragment of the classical propositional calculus. Mihailescu's decision criterion has been formulated independently of McKinsey's (cf., e.g., [1]).

Mihailescu also proves consistency and independence of the axiom system (I) and (XXIII) using the matrix method.

9. Rasiowa's axiomatization of the calculus with equivalence and non-equivalence

In the article [8] written in 1937 Mihailescu mentions the fact that the propositional calculus with the connectives E and E' is a deductively incomplete fragment of the classical propositional calculus as, e.g., the formula $E'pp$ cannot be either proved or refuted. In 1947 Helena Rasiowa described the calculus with the connectives E and E' in the paper [12]. On the basis of Leśniewski's paper [3] the author proved that the axiom system consisting of the shortest axiom of Łukasiewicz (XI) and of the following axiom

(XXIV) $EEpqEE'rqE'pr$

is consistent, independent, complete and decidable. The decision criterion given by Rasiowa decides that any propositional formula built up by means of the connectives E and E' may be proved in the propositional calculus with the connectives E and E' if and only if both every variable and the connective E' occur an even number of times in it. Obviously, the connective E occurs in the discussed formula an odd number of times.

10. Stone's criterion for the calculus with equivalence, negation and disjunction

In the above mentioned paper [16] Stone also discusses a functionally complete system of the classical propositional calculus with the connectives E, N and the disjunction A and settles for it the following criterion of decidability.

LEMMA 11. *The set of all formulas built up of variables by means of the connectives* E, N *and* A *treated as the operations characterized by the axioms* (A1) *and* (A2) *and the following axioms:*

(XXV) $EApAqrAApqr$, $EAEpqrEArpArq$, $EAppp$, $EAEpNpqq$

with the relation of identity $\equiv$ *defined by the condition*

(i) $x \equiv y$ *if and only if* Exy *can be deduced from the axioms* (A1), (A2) *and* (XXV) *by means of the rules of detachment, substitution and the rule defined by the scheme*

$$\frac{x}{Axy}$$

for any formulas x *and* y *built up of variables and the connectives* E, N *and* A,

forms a Boolean ring.

Functionally complete enrichments of the equivalential propositional calculus were described in a series of papers by E. Gh. Mihailescu published in the Rumanian periodicals *Revue Roumaine des Sciences Sociales. Série de Philosophie et Logique* and Acta Logica after Second World War. This subject was also investigated by Jerzy Słupecki in [14] in 1953 and by Czesław Lejewski in [2] in 1968.

11. Deduction theorems for the equivalential and equivalential-negational calculus

To finish this review we shall also present the deduction theorems valid in the equivalential and in the equivalential-negational calculus respectively, and formulated in the paper [17].

Schemes:

(XXVI) $\frac{x,\ Exy,}{y}\quad\frac{x,\ Eyx}{y}$

(where x and y are any equivalential formulas) define two rules of detachment valid in the equivalential propositional calculus.

Let A be any fixed set of the equivalential and equivalential-negational formulas, respectively. Let us consider the following conditions:

(XXVII) For any set of the equivalential formulas X and for any equivalential formulas x, y if the formula x can be deduced from the set of formulas $A+X+\{y\}$ by means of the rules (XXVI), then either the formula x or the formula Eyx can be deduced from the set of formulas $A+X$ by means of the rules (XXVI)

(XXVIII) For any set of the equivalential formulas X and for any equivalential formulas x, y if the formula x can be deduced from the set of formulas $A+X+\{y\}$ by means of the rules (XXVI), then either the formula Eyx or the formula $EyEyx$ can be deduced from the set of formulas $A+X$ by means of the rules (XXVI).

The conditions (XXVII) and (XXVIII) are called the direct deduction theorems for the equivalential propositional calculus. The following lemma holds:

LEMMA 12. *Holding of any of the conditions* (XXVII) *or* (XXVIII) *is equivalent to the fact that the set* A *coincides with an axiom system of the set of all theorems of the equivalential fragment of the classical propositional calculus.*

The next condition is called the indirect deduction theorem for the equivalential-negational propositional calculus.

(XXIX) For any set of the equivalential-negational formulas X and for any equivalential-negational formulas x, y if formulas x, Nx can be deduced from the set of the formulas $A+X+\{Ny\}$ by means of the rules (XXVI), then either the formula y or the formula $ENyy$ can be deduced from the set of the formulas $A+X$ by means of the rules (XXVI)

The following holds:

LEMMA 13. *Holding of the condition* (XXIX) *and of the conditions* (XXVII) *or* (XXVIII), *this time formulated for equivalential-negational formulas, is equivalent to the fact that the set* A *coincides with an axiom system of the set of all theorems of the equivalential-negational fragment of the classical propositional calculus.*

Sufficiency proofs in lemmas 12 and 13 are based upon Leśniewski's criterion. Necessity proofs are carried out by means of induction on the definition of deducibility.

The cited theorems of deduction (XXVII), (XXVIII) and (XXIX) facilitate and simplify formalized proofs of the theorems in the equivalential and in the equivalential-negational calculus, respectively.

References

[1] CHURCH A.: *Introduction to mathematical logic.* Vol. I. Princeton 1956.

[2] LEJEWSKI Cz.: *A propositional calculus in which three mutually undefinable functors are used as primitive terms.* Studia Logica, 22(1968), 17–46.

[3] LEŚNIEWSKI S.: *Grundzüge eines neuen Systems der Grundlagen der Mathematik·* Einleitung und § 1–11.
Fundamenta Mathematicae, 14(1929), 1–81.

[4] LEŚNIEWSKI S.: *Einleitende Bemerkungen zur Fortsetzung meiner Mitteilung u. d. T. „Grundzüge eines neuen Systems der Grundlagen der Mathematik".*
Collectanea Logica, 1(1939), 1–60, Warszawa.

[5] ŁUKASIEWICZ J.: *Elementy logiki matematycznej.* Skrypt autoryzowany opracował M. Presburger.
Nakładem Komisji Wydawniczej Koła Matematyczno-Fizycznego Słuchaczów Uniwersytetu Warszawskiego, 1929.

[6] ŁUKASIEWICZ J.: *Der Äquivalenzenkalkül.*
Collectanea Logica, 1(1939), 145–169, Warszawa.

[7] MEREDITH C. A., PRIOR A. N.: *Notes on the axiomatics of the propositional calculus.*
Notre Dame Journal of Formal Logic, 4(1963), 171–187.

[8] MIHAILESCU E. GH.: *Recherches sur un sous-système du calcul de propositions.*
Annales Scientifiques de l'Université de Jassy. Partie I, 23(1937), 106–124.

[9] MIHAILESCU E. GH.: *Recherches sur la négation et l'équivalence dans le calcul des propositions.*
Ibid., 23(1937), 369–408.

[11] QUINE W. V.: *Review of the paper of M. H. Stone, „Note on formal logic".* American Journal of Mathematics, 59(1937), 506–514.
Journal of Symbolic Logic, 2(1937), 174–175.

[11] QUINE W. V.: *Mathematical logic.*
New York 1940.

[12] RASIOWA H.: *Axiomatisation d'un système partiel de la théorie de la déduction.*
Comptes Rendus des Séances de la Société des Sciences et des Lettres de Varsovie, Classe III, 40(1947), 22–37.

[13] RASIOWA H.: *O pewnym fragmencie implikacyjnego rachunku zdań.*
Studia Logica, 3(1955), 208–222.

[14] SŁUPECKI J.: *Über die Regeln des Aussagenkalküls.*
Studia Logica, 1(1953), 19–43.

[15] SOBOCIŃSKI B.: *Z badań nad teorią dedukcji.*
Przegląd Filozoficzny, 35(1932), 3–193.

[16] STONE M. H.: *Note on formal logic.*
American Journal of Mathematics, 59(1937), 506–514.

[17] SURMA S. J.: *Method of natural deduction in equivalential and equivalential-negational propositional calculus.*
Universitas Iagellonica Acta Scientiarum Litterarumque, CCLXV, Schedae Logicae, 6(1971), 55–68.

[18] TAJTELBAUM A.: *Sur le terme primitif de la logistique.*
Fundamenta Mathematicae, 4(1923), 196–200.

[19] WAJSBERG M.: *Ein neues Axiom der Aussagenkalküls in der Symbolik von Sheffer.*
Monatshefte für Mathematik und Physik, 39(1932), 259.

[20] WAJSBERG M.: *Metalogische Beiträge.*
Wiadomości Matematyczne, 43(1937), 1–38, and 47(1939), 119–139.

[21] WHITEHEAD A. N., RUSSELL B.: *Principia Mathematica.* Vol. I.
Cambridge, England, 1910.

STANISŁAW J. SURMA

A UNIFORM METHOD OF PROOF OF THE COMPLETENESS THEOREM FOR THE EQUIVALENTIAL PROPOSITIONAL CALCULUS AND FOR SOME OF ITS EXTENSIONS [1]

Introduction

The present paper is concerned with the adjustment of the method of proving the completeness of the classical propositional calculus which was described by G. Asser in [1], for the completeness proof of the equivalential fragment of this calculus. This method will be adopted also for the calculi based on equivalence and negation and for those based on equivalence and non-equivalence, i.e., exclusive disjunction.

The method described by G. Asser amounts in fact to a slight modification of the method of J. Łoś (cf. [41]) which is in turn an adaptation of the well-known L. Henkin's method of proving the completeness of the classical first order predicate calculus. Asser's method is based upon Lindenbaum's theorem on deductively complete supersystems as well as upon Tarski-Herbrand's deduction theorem for the implication connective. In the present paper we utilize the deduction theorem for the equivalence connective stated and proved in [7].

In respect of the number of the theorems of the propositional calculus in question used in the proof of the completeness of this calculus the presented method seems to be the simplest of the hitherto known methods of proving the completeness of the purely equivalential propositional calculus and of some of its extensions (cf. [8]). The completeness proof for the purely equivalential propositional calculus which is due to S. Leśniewski (cf. [2]), utilizes seventy nine auxiliary theorems while a simpli-

---

[1] This paper is an extended version of a lecture given by the author on April 24, 1971 to the XVIIth Conference of History of Logic organized in Cracow by the Section of Logic, Polish Academy of Sciences, together with the Department of Logic of the Jagiellonian University.

fication of the Leśniewski's proof given by J. Łukasiewicz in [3] makes use of fourty eight auxiliary theorems. The completeness proof for the propositional calculus based on equivalence and negation given by E. Mihailescu utilizes two hundred and twenty auxiliary theorems. Again, the completeness proof for the propositional calculus based on equivalence and non-equivalence given, by H. Rasiowa in [6] makes use of sixty auxiliary theorems. The completeness proofs for the quoted propositional calculi which are given in the present paper are utilizing, respectively, twenty, thirty four and thirty one auxiliary theorems.

We denote the propositional variables by the letters

$$p, q, r, \ldots$$

The connectives of equivalence, negation and non-equivalence are denoted, respectively, by

$$E, N, E'.$$

Arbitrary propositional formulas construed of the formulas x, y and of the connectives are symbolized by

$$Exy, Nx, E'xy.$$

1. The completeness proof for the equivalential propositional calculus

In the present chapter we shall confine ourselves to considering only the propositional formulas construed exclusively of propositional variables and of the equivalence connective E.

The set of all the theorems of the equivalential fragment of the classical propositional calculus is defined now to be the smallest set comprising the formulas:

(A1) $EEExzEyxEzy$

and

(A2) $EExEyzEExyz$

which is at the same time, closed under the rule defined by the following scheme

(R1) $Exy, x/y$

which is called the rule of detachment.

The axioms (A1) and (A2) are due to S. Leśniewski who was the first to give an axiomatization of the equivalential propositional calculus (cf. [2]).

By the rule (R1) we deduce from the axioms (A1) and (A2) twenty theorems which we shall utilize in the sequel.

(T1) $EEyzEzy$
A1 $[x/z, y/Ezy, z/Eyz]$, A2 $[x/z]$, R1.

The inscription above means that the theorem which has the number (T1) was obtained by applying the rule (R1) to the axioms (A1) and (A2); at the same time the formulas x, y, z in the axiom (A1) being replaced by the formulas z, Ezy, and Eyz respectively, while in the axiom (A2) the formula x was replaced by the formula z.

It follows from theorem (T1) that the set of theorems of the purely equivalential propositional calculus is closed under the rule of deduction given by the scheme

(R2) $Exy, y/x$

which will be also called the rule of detachment.

(T2) $EEEyxExyEyExExy$
A2 $[x/y, y/x, z/Exy]$, T1 $[y/EyExExy, z/EEyxExy]$, R1

(T3) $EyExExy$
T1 $[z/x]$, T2, R1

(T4) Ezz
A1 $[x/y, y/z]$, T1, R1

(T5) $EEzyEExzEyx$
A1, T1 $[y/EExzEyx, z/Ezy]$, R1

(T6) $EExEyzEEzyx$
T1, T5 $[y/Ezy, z/Eyz]$, R1

(T7) $EEzExyExEyz$
A2, T6 $[x/ExEyz, y/Exy]$, R1

(T8) $EEyzEExzEyx$
A1, T6 $[x/EExzEyx, y/z, z/y]$, R1

(T9) $EExzEEyxEyz$
T7 $[x/Exz, y/Eyx, z/Eyz]$, T8, R1

(T10) $EEyxEEyzExz$
T7 $[x/Exy, y/Eyz, z/Exz]$, T9, R1

(T11) $EEEyzuEEzyu$
T1, T10 $[x/Ezy, y/Eyz, z/u]$, T1, R1

(T12) *EExyEEyzExz*
T10, T11 [z/x, $u/EEyzExz$], R1

(T13) *EEEEyzExzuEExyu*
T12 [x/Exy, $y/EEyzExz$, z/u], T12, R1

(T14) *EExEyzEEuyExEuz*
T13 [y/Eyz, z/Euz, $u/EEuyExEuz$],
T13 [x/u, $u/ExEuz$], R1

(T15) *EExEEyzuExEEzyu*
T9 [$x/EEyzu$, y/x, $z/EEzyu$], T11, R1

(T16) *EExEyzEEyxz*
A2, T15 [$x/ExEyz$, y/x, z/y, u/z], R1

LEMMA 1. *The set of theorems of the purely equivalential propositional calculus is closed under the rule of extensionality, i.e., under the rule given by the scheme*

(R3) Exy/Ezz^x_y

where z^x_y *symbolizes the formula obtained by replacing subformula* x *in the formula* z *by formula* y. *To put it more precisely, the rule* (R3) *is derivable by means of the rule* (R1), *on the grounds of the theorems* (T1), (T4) *and* (T10).

Proof. Assume that the following formula is valid

(1) Exy.

We shall prove now that the following formula is likewise valid

(2) Ezz^x_y

by induction on the construction of the formula z. If $z = p = x$ then $z^x_y = y$, and according to (1) we have (2). If on the contrary, $z = p \neq x$, then $z^x_y = p$, and according to theorem (T4) we also obtain (2). Let

(3) $z = Euw$

and assume inductively that lemma 1 holds for the formulas u, w. Then in view of (1), there hold the following formulas

(4) Euu^x_y

and

(5) Eww^x_y.

By virtue of theorem (T14) the following formula is valid

(6) $EEEuwEuw^x_yEEu^x_yuEEuwEu^x_yw^x_y$.

According to theorem (T9)

(7) $EEww^x_yEEuwEuw^x_y$.

Hence and from (5)

(8) $EEuwEuw^x_y$.

Hence in turn, and from (6)

(9) $EEu^x_yuEEuwEu^x_yw^x_y$.

From (4) on the other hand, and from theorem (T1) we obtain

(10) Eu^x_yu.

Hence and from (9)

(11) $EEuwEu^x_yw^x_y$.

Since

(12) $Eu^x_yw^x_y = (Euw)^x_y$

then in view of (3) we finally obtain from (11) the formula

(13) Ezz^x_y

which completes the proof of the lemma.

(T17) $EExzEEyzEyx$
T1 $[y/Eyx, z/Eyz]$, T9, R3

(T18) $EExEyzEzExy$
A2, T1 $[y/Exy]$, R3

(T19) $EExEyzEyExz$
A2 $[x/y, y/x]$, T16, R3

(T20) $EExEyzEExzy$
T1 $[z/Exz]$, T19, R3

Let $Cn(R\colon A+X)$ symbolize the set of all propositional formulas deducible from the set $A+X$ of the propositional formulas by means of the set of rules R. In the sequel we shall make use of the following modification of the classical deduction theorem which was given in [7]:

($W_{R,A}$) For arbitrary propositional formulas x, y and for an arbitrary set of propositional formulas X, if $x \in Cn(R;\ A+\{y\})$, then $x \in Cn(R;\ A+X)$, or $Eyx \in Cn(R;\ A+X)$

LEMMA 2. (*cf.* [7]). *If* $R = \{R1, R2\}$ *and* $A = \{A2, T1, T4, T9, T17, T18, T19, T20\}$ *or if* $R = \{R1\}$ *and* $A = \{A2, T4, T9, T19\}$, *then* ($W_{R,A}$).

Let e be a binary operation in the set $\{0, 1\}$ such that

$$e(x, y) \text{ if and only if } x = y$$

for all x, $y \in \{0, 1\}$. A propositional formula x is a tautology of the equivalential propositional calculus if for an arbitrary mapping f of the propositional variables into the set $\{0, 1\}$, and for an arbitrary mapping h, the fact that h satisfies the conditions:

(i) $h(p) = f(p)$

for an arbitrary propositional variable p, and

(ii) $h(Eyz) = e(h(y), h(z))$

for all propositional formulas x, y, entails that $h(x) = 1$.

LEMMA 3. *Every theorem of the equivalential propositional calculus is a tautology of this calculus.*

Let x be an arbitrary propositional formula. Let $\mathfrak{L}(x)$ symbolize the family of the sets of the propositional formulas of the form $L(x)$ such that

(i) $L(x)$ is an extension of the set of all the theorems of the equivalential propositional calculus, closed under the rule of detachment (R1),

(ii) $x \notin L(x)$

and

(iii) if $y \notin L(x)$, then $x \in Cn\,(\mathrm{R1};\ L(x) + \{y\})$ for an arbitrary propositional formula y.

LEMMA 4. *For an arbitrary propositional formula x, either x is a theorem of the equivalential propositional calculus or the family $\mathfrak{L}(x)$ is non-empty.*

Proof. Let x not be a theorem of the equivalential propositional calculus. Let us make a sequence of all the propositional formulas

(1) $x_1, x_2, x_3, \ldots$

We define the monotonic sequence of the propositional formulas $\{X_i\}_{i \in \omega}$ in the following way:

(2) X_0 is the set of all theorems of the equivalential propositional calculus

(3) $X_{n+1} = X_n$, if $x \in Cn\,(\mathrm{R1};\ X_n + \{x_n\})$

(4) $X_{n+1} = X_n + \{x_n\}$, if $x \notin Cn\,(\mathrm{R1};\ X_n + \{x_n\})$.

We put

(5) $L(x) = \bigcup \{X_i\colon\ i \in \omega\}$.

It is easy to prove that the set $L(x)$ has the properties (i)–(iii) stated in the definition of the family $\mathfrak{L}(x)$.

This lemma is essentially a modification, viz. in the point (iii) of the definition of the family $\mathfrak{L}(x)$, of the well known Lindenbaum's theorem on complete supersystems.

LEMMA 5. *For arbitrary propositional formulas* x, y, z, *and for an arbitrary* $L(x) \in \mathfrak{L}(x)$ *the following conditions are equivalent*:

(a) $Eyz \in L(x)$

and

(b) $y \in L(x)$ *if and only if* $z \in L(x)$.

Proof. Let $y \in L(x)$. Since the set $L(x)$ is closed under the rule (R1), then in view of (a), we have that $z \in L(x)$. Conversely, let $z \in L(x)$. The condition (a) and the theorem (T1) yield that $Ezy \in L(x)$; while hence, by means of the rule (R1) it follows that $y \in L(x)$. Thus, the condition (a) entails the condition (b). Assume in turn, that

(1) $y \in L(x)$

and

(2) $z \in L(x)$.

The set $L(x)$ is an extension of the set of all theorems of the classical equivalential propositional calculus closed under the rule (R1). Hence and from the theorem (T3) it follows that

(3) $EzEyEyz \in L(x)$.

And hence, in view of (2) and (1) it follows according to (R1), that

(4) $Eyz \in L(x)$.

Assume, on the contrary, that

(5) $y \notin L(x)$

and

(6) $z \notin L(x)$.

The set $L(x)$ satisfies the conditions (i)–(iii) of the definition of the family $\mathfrak{L}(x)$. Hence and from (5) it follows that

(7) $x \in Cn\,(\mathrm{R1};\ L(x)+\{y\})$.

Applying lemma 2 to (7) we obtain that

(8) $x \in L(x)$, or $Eyx \in L(x)$.

From (8) and from the condition (ii) of the definition of the family $\mathfrak{L}(x)$ it follows that

(9) $Eyx \in L(x)$.

By analogous reasoning from (6) it follows that

(10) $Ezx \in L(x)$.

In view of the theorem (T8) we obtain that

(11) $EEyxEEzxEyz \in L(x)$.

This together with (9) and (10) yields by means of the rule (R1) that

(12) $Eyz \in L(x)$

which completes the proof of this lemma.

We shall now pass to proving the lemma which is essential for this chapter.

LEMMA 6. *Every tautology of the equivalential propositional calculus is a theorem of this calculus.*

Proof. Assume that

(1) x is not a theorem of the equivalential propositional calculus.

Hence in view of lemma 4, it follows that there exists a set $L(x)$ which fulfils the conditions (i)–(iii) of the definition of the family $\mathfrak{L}(x)$. We define the mapping f of the propositional variables into the set $\{0, 1\}$, putting

(2) $f(p) = 1$ if and only if $p \in L(x)$

for all propositional variables p. By induction on construing an arbitrary formula y we shall prove that

(3) $h(y) = 1$ if and only if $y \in L(x)$

where h is an extension of f and it fulfils the conditions (i)–(iii) of the definition of the tautology of the equivalential propositional calculus. If $y = p$, then the step (3) is coincident with the step (2). Let then

(4) $y = Ezu$.

Assume inductively that

(5) $h(z) = 1$ if and only if $z \in L(x)$

and

(6) $h(u) = 1$ if and only if $u \in L(x)$.

Assume that

(7) $h(Ezu) = 1$.

Then

(8) $e(h(z), h(u)) = 1$

and as a consequence

(9) $h(z) = h(u)$.

Let

(10.1) $h(z) = h(u) = 1$.

Hence and from (5) and (6)

(10.2) $z,\ u \in L(x)$

and this together with lemma 5 yields

(10.3) $Ezu \in L(x)$.

Then

(10) if $h(z) = h(u) = 1$, then $Ezu \in L(x)$.

Let now

(11.1) $h(z) = h(u) = 0$.

This, (5) and (6) entail

(11.2) $z \notin L(x)$ and $u \notin L(x)$

and hence, and from lemma 5

(11.3) $Ezu \in L(x)$.

Thus

(11) if $h(z) = h(u) = 0$, then $Ezu \in L(x)$.

From (10) and (11) it follows that

(12) $Ezu \in L(x)$.

We have thus proved that

(13) if $h(Ezu) = 1$, then $Ezu \in L(x)$.

The implication converse to (13) follows analogously from the inductive assumptions (5) and (6), and form the definitions of the functions f and h. Thus the step (3) is proved. Further, according to the definition of the family $\mathfrak{L}(x)$ we have that

(14) $x \notin L(x)$.

Hence and from (3) it follows that

(15) $h(x) = 0$

and this finally yields that x is not a tautology of the equivalential propositional calculus.

Lemmas 3 and 6 immediately entail the completeness of the here described equivalential fragment of the classical propositional calculus.

2. The completeness proof for the propositional calculus based on the equivalence and negation connectives

The set of all theorems of the propositional calculus with equivalence and negation is defined as the smallest set including (A1), (A2) and

(A3) $EExyENxNy$

and closed under the rule (R1).

LEMMA 7. *The set of all theorems of the propositional calculus based on equivalence and negation connectives is closed under the rule extensionality* (R3).

Proof. In view of lemma 1 it should only be shown that if Euu^x_y is a theorem of the calculus under consideration, then also $ENuNu^x_y$ is a theorem of this calculus. Let

(1) $z = Nu.$

Assume inductively that the lemma holds for the formula u, i.e., that the following formula holds:

(2) $Euu^x_y.$

According to axiom (A3)

(3) $EEuu^x_yENuNu^x_y.$

From (2) and (3), according to the rule (R1) it follows that

(4) $ENuNu^x_y$

and therefrom and from (1) we finally obtain the formula

(5) Ezz^x_y

which completes the proof as $Nu^x_y = (Nu)^x_y$.

From the axioms (A1), (A2) and (A3) we shall deduce now the fourteen subsequent theorems making use of the rules (R1), (R2) and (R3).

(T21) $EENxNyExy$
A3, T1 $[y/Exy, z/ENxNy]$, R1

(T22) $ENyENxEyx$
A2 $[x/Ny, y/Nx, z/Eyx]$, T21 $[x/y, y/x]$, R2

(T23) $EENxEyxENxExy$
T1 $[z/x]$, T4 $[z/ENxEyx]$, R3

(T24) $ENyENxExy$
T22, T23, R3

(T25) $ENyEENxxy$
A2 $[x/Ny, y/Nx, z/Exy]$, T24, R1

(T26) $ENyEyENxx$
T1 $[y/ENxx, z/y]$, T25, R3

(T27) $EEENyNxExxExEENyNxx$
A2 $[x/ENyNx, y/x, z/x]$, T1 $[y/EENyNxx, z/x]$, R3

(T28) $EEENyNxExxExENyENxx$
A2 $[x/Ny, y/Nx, z/x]$, T27, R3

(T29) $EEENyNxExxEExNyENxx$
A2 $[y/Ny, z/ENxx]$, T28, R3

(T30) $EEENyNxExxEENyxENxx$
T1 $[y/x, z/Ny]$, T29, R3

(T31) $EEEyxENyNxEEEyxENyNxExx$
T3 $[x/EEyxENyNx, y/$T4$]$, T4 $[z/x]$, R1

(T32) $EEEyxENyNxExx$
T31, T21 $[x/y, y/x]$, R1

(T33) $EEyxEENyNxExx$
T2 $[x/Eyx, y/ENyNx, z/Exx]$, T32, R2

(T34) $EEyxEENyxENxx$
T30, T33, R3

We shall now supplement the definition of the concept of tautology from the previous chapter. Let n be a unary operation in the set $\{0, 1\}$, and such that $n(x) = 1-x$ for any $x \in \{0, 1\}$. The propositional formula x is a tautology of the propositional calculus with equivalence and negation connectives if for any mapping f of the propositional variables into the set $\{0, 1\}$, and for any mapping h, the fact that h satisfies the conditions

(i) $h(p) = f(p)$

for any propositional variable p,

(ii) $h(Eyz) = e(h(y), h(z))$

and

(iii) $h(Ny) = n(h(y))$

for any propositional formulas x and y, entails that $h(x) = 1$.

LEMMA 8. *Every theorem of the propositional calculus with the equivalence and negation connectives is a tautology of this calculus.*

We shall modify the definition of the family $\mathfrak{L}(x)$ of the previous chapter as follows. Let x be an arbitrary propositional formula. Let us symbolize by $\mathfrak{L}_N(x)$ the family of the sets of the propositional formulas of the form $L(x)$, and such that

(i) $L(x)$ is an extension of the set of all theorems of the propositional calculus with the equivalence and negation connectives and closed under the rule (R1)

(ii) $x \notin L(x)$

(iii) if $y \notin L(x)$, then $x \in Cn\,(\mathrm{R1};\ L(x)+\{y\})$, for an arbitrary propositional formula y

and

(iv) $Nx \in L(x)$.

LEMMA 9. *For an arbitrary propositional formula x, either x is a theorem of the propositional calculus with the equivalence and negation connectives, or the family $\mathfrak{L}_N(x)$ is non-empty.*

The proof of this lemma is analogous to that of lemma 4; it only differs from the latter in the definition of the set X_0. Presently we require that $X_0 = X_0' + \{Nx\}$, where X_0' is the set of all theorems of the propositional calculus with the equivalence and negation connectives. Obviously, x would not be deducible from the set X_0. Otherwise it would be that

(1) $x \in Cn\,(\mathrm{R1};\ X_0' + \{Nx\})$.

This and lemma 2 entail that

(2) $x \in X_0'$, or $ENxx \in X_0'$.

Therefrom and from the assumption of the lemma which is being proved we deduce

(3) $ENxx \in X_0'$.

But by virtue of lemma 8 it would give

(4) $ENxx$ is a tautology of the propositional calculus with the equivalence and negation connectives which is not possible.

LEMMA 10. *For any propositional formulas x, y, z and for arbitrary $L(x) \in \mathfrak{L}_N(x)$*

(a) $Eyz \in L(x)$ *if and only if* $y \in L(x)$ *if and only if* $z \in L(x)$

(b) $Ny \in L(x)$ *if and only if* $y \notin L(x)$.

Proof of (a) is just a repetition of the proof of lemma 5. We have only to prove (b) now. Assume that

(1) $Ny \in L(x)$.

In view of the theorem (T26)

(2) $ENyEyENxx \in L(x)$.

From (1) and (2) it follows that

(3) $EyENxx \in L(x)$.

By virtue of (a) we obtain from (3) that

(4) $y \in L(x)$ if and only if $ENxx \in L(x)$.

But according to the points (ii) and (iv) of the definition of $\mathfrak{L}_N(X)$ we have

(5) $ENxx \notin L(x)$

Therefrom and from (4) it follows that

(6) $y \notin L(x)$.

Thus the step (1) entails the step (6). Assume in turn (6) and suppose that

(7) $Ny \notin L(x)$.

From this and from the definition of the family $\mathfrak{L}_N(x)$ it follows that

(8) $x \in Cn(\mathrm{R}1;\ L(x)+\{y\})$

and hence in view of lemma 2 we obtain that

(9) $Eyx \in L(x)$.

By analogous reasoning we deduce from (7) that

(10) $ENyx \in L(x)$.

Observe, however, that in view of the theorem (T34)

(11) $EEyxEENyxENxx \in L(x)$.

The steps (9), (10) and (11) entail that

(12) $ENxx \in L(x)$

which is not possible. Thus also the step (6) entails the step (1).

LEMMA 11. *Every tautology of the propositional calculus with equivalence and negation connectives is a theorem of this calculus.*

Proof. We assume that

(1) x is not a theorem of the propositional calculus with the equivalence and negation connectives

On the grounds of lemma 9 we deduce that there exists a set $L(x)$ satisfying the conditions (i)–(iv) of the definition of the family $\mathfrak{L}_N(x)$. We define a mapping f of the propositional variables into the set $\{0,1\}$, putting

(2) $f(p) = 1$ if and only if $p \in L(x)$

for any propositional formula p. By induction on the construction of an arbitrary propositional formula y we demonstrate that

(3) $h(y) = 1$ if and only if $y \in L(x)$

where h is an extension of f and it satisfies the conditions (i)–(iii) of the definition of a tautology of the propositional calculus with the equivalence and negation connectives. In view of the definition of the family $\mathfrak{L}_N(x)$

(4) $x \notin L(x)$.

This and (3) yield

(5) $h(x) = 0$

and hence it finally follows that x is not a tautology.

From lemmas 8 and 11 there follows the completeness of the propositional calculus with the equivalence and negation connectives.

3. The completeness proof for the propositional calculus based on the equivalence and non-equivalence connectives

The set of all theorems of the propositional calculus with the equivalence and non-equivalence is defined to be the smallest set including (A1), (A2) and

(A4) $EExyEEzuEE'xzE'yu$

and closed under the rule (R1).

LEMMA 12. *The set of all theorems of the propositional calculus with the equivalence and non-equivalence connectives is closed under the rule of extensionality* (R3).

From the axioms (A1), (A2), and (A4) we deduce by means of the rules (R1), (R2) and (R3) the subsequent eleven theorems.

(T35) $EEExzEyzEE'xyE'zz$
A2 $[x/Exz, y/Eyz, z/EE'xyE'zz]$, A4 $[y/z, z/y, u/z]$, R1

(T36) $EExyEE'xyE'zz$
T10 $[x/y, y/x]$, T35, R3

(T37) $EE'xyEExyE'zz$
T19 $[x/Exy, y/E'xy, z/E'zz]$, T36, R1

(T38) $EE'xyEExEE'zzy$
T20 $[y/E'zz, z/y]$, T37, R3

(T39) $EEzxEEyzEE'zzE'xy$
A4 $[x/z, y/x, u/y]$, T1, R3

(T40) $ExEzEEyzEE'zzE'xy$
T16 $[y/z, z/EEyzEE'zzE'xy]$, T39, R2

(T41) $ExEEyzEzEE'zzE'xy$
T19 $[x/z, y/Eyz, z/EE'zzE'xy]$, T40, R3

(T42) $ExEEyzEEE'zzzE'xy$
T16 $[x/z, y/E'zz, z/E'xy]$, T41, R3

(T43) $EEExyEyxEE'xyE'yx$
A2 $[x/Exy, y/Eyx, z/EE'xyE'yx]$, A4 $[z/y, u/x]$, R1

(T44) $EE'xyE'yx$
T1 $[y/x, z/y]$, T43, R1

(T45) $ExEEyzEEE'zzzE'yx$
T42, T44, R3

We shall supplement now the definition of tautology given in the first chapter. Let e' be a binary operation in the set $\{0,1\}$, such that $e'(x,y) = 1$ if and only if $x \neq y$ for any x, $y \in \{0,1\}$. Propositional formula x is a tautology of the propositional calculus with the equivalence and non-equivalence connectives if for any mapping f of the propositional variables into the set $\{0,1\}$ and for any h being an extension of the mapping f and conforming with the operations e and e' we have the equality $h(x) = 1$ valid.

LEMMA 13. *Every theorem of the propositional calculus with the equivalence and non-equivalence connectives is a tautology of this calculus.*

Let us modify now the definition of the family $\mathfrak{L}(x)$ which was given in the first chapter. Let x be an arbitrary propositional formula. Let us symbolize by $\mathfrak{L}_{E'}(x)$ the family of sets of the propositional formulas of the form $L(x)$ and such that

(i) $L(x)$ is an extension of the set of all theorems of the propositional calculus with the equivalence and non-equivalence connectives and closed under R1

(ii) $x \notin L(x)$, and $E'xx \notin L(x)$

and

(iii) if $y \notin L(x)$, then $x \in Cn(\text{R1}; L(x) + \{y\})$, for any propositional formula y.

LEMMA 14. *For any propositional formula x, either x is a theorem of the propositional calculus with the equivalence and non-equivalence connectives, or the family $\mathfrak{L}_{E'}(x)$ is non-empty.*

LEMMA 15. *For any propositional formulas x, y, z and for any* $L(x) \in \mathfrak{L}_{E'}(x)$

(a) $Eyz \in L(x)$ *if and only if* $y \in L(x)$ *if and only if* $z \in L(x)$

(b) $E'yz \in L(x)$ *if and only if* $y \in L(x)$ *if and only if* $z \notin L(x)$.

Proof of (a) amounts to rewriting the proof of lemma 5. Proof of (b) is analogous to the proof of lemma 10 and utilizes lemmas 2 and 5, the definition of the family $\mathfrak{L}_{E'}(x)$ and the theorems (T37), (T38), (T42) and (T45).

LEMMA 16. *Every tautology of the propositional calculus with the equivalence and non-equivalence connectives is a theorem of this calculus.*

Proof of this lemma is analogous to that of lemma 11 and is based on lemma 15.

Lemmas 13 and 16 yield the completeness of the described propositional calculus with equivalence and non-equivalence connectives.

4. A simple generalization of the described method

Let F_k be a k-ary connective which satisfies the axiom

(F_k) $EEx_1y_1EEx_2y_2 \dots EEx_ky_kEF_kx_1x_2 \dots x_kF_ky_1y_2 \dots y_k$

We define the concept of tautology of the propositional calculus with the equivalence connective and with the F_k connective. Viz., let f_k be a k-ary operation in the set $\{0, 1\}$ defined inductively as follows

(i) $f_1(x_1) = 1 - x_1$

(ii) $f_{k+1}(x_1, x_2, \dots, x_{k+1}) = e(f_k(x_1, x_2, \dots, x_k), x_{k+1})$

for $k \geqslant 1$, and for any $x_1, x_2, \dots, x_{k+1} \in \{0, 1\}$. Let f be an arbitrary mapping of the propositional variables into the set $\{0, 1\}$ and let h be an arbitrary mapping such that for any propositional variables p

(i) $h(p) = f(p)$

and for any propositional formulas $x, y, x_1, x_2, \dots, x_k$

(ii) $h(Exy) = e(h(x), h(y))$

and

(iii) $h(F_kx_1x_2 \dots x_k) = f_k(h(x_1), h(x_2), \dots, h(x_k))$.

Then the propositional formula x is a tautology if and only if $h(x) = 1$

It can be proved now that every theorem of the propositional calculus with the connectives E and F_k is a tautology of this calculus.

For any propositional formula x we define the family $\mathfrak{L}_{F_k}(x)$ of the sets of the form $L(x)$, as follows:

(i) $L(x)$ is an extension of the set of all theorems of the propositional calculus with the connectives E and F_k, and closed under the rule (R1),

(ii) $x \notin L(x)$,

(iii) if k is an even number, then $F_k\underbrace{xx \dots x}_{k \text{ times}} \notin L(x)$,

(iv) if k is an odd number, then $F_k\underbrace{xx \dots x}_{k \text{ times}} \in L(x)$

and

(v) if $y \notin L(x)$, then $x \in Cn(\mathrm{R1}; L(x) + \{y\})$, for any propositional formula y.

It can be proved that for any propositional formula x, either x is a theorem of the propositional calculus with the connectives E and F_k, or the family $\mathfrak{L}_{F_k}(x)$ is non-empty.

Further, it can be demonstrated that for arbitrary propositional formulas $x_1, x_2, \ldots, x_k$ and for an arbitrary $L(x) \in \mathfrak{L}_{F_k}(x)$

(i) if $k = 1$, then $F_k x_1 x_2 \ldots x_k = F_1 x_1$ and in this case $F_1 x_1 \in L(x)$ if and only if $x_1 \notin L(x)$,

(ii) if $k > 1$, then $F_k x_1 x_2 \ldots x_k \in L(x)$ if and only if $F_{k-1} x_1 x_2 \ldots x_{k-1} \in L(x)$ if and only if $x_k \in L(x)$.

Finally it can be proved that every tautology of the propositional calculus with the connectives E and F_k is a theorem of this calculus.

References

[1] ASSER G.: *Einführung in die mathematische Logik*. Teil I.
Leipzig 1959.

[2] LEŚNIEWSKI S.: *Grundzüge eines neuen Systems der Grundlagen der Mathematik*. Einleitung und § 1–11.
Fundamenta Mathematicae, 14 (1929), 1–81.

[3] LEŚNIEWSKI S.: *Einleitende Bemerkungen zur Fortsetzung meiner Mitteilung u. d. T. „Grundlagen eines neuen Systems der Grundlagen der Mathematik"*.
Collectanea Logica, 1 (1939), 1–60, Warszawa.

[4] ŁOŚ J.: *An algebraic proof of completeness for the two-valued propositional calculus*.
Colloquium Mathematicum, 2 (1951), 236–240.

[5] MIHAILESCU E. GH.: *Recherches sur un sous-système du calcul de propositions*.
Annales Scientifiques de l'Université de Jassy. Partie I, 23 (1937), 106–124.

[6] RASIOWA H.: *Axiomatisation d'un système partiel de la théorie de la déduction*.
Comptes Rendus des Séances de la Société des Sciences et des Lettres de Varsovie, Classe III, 40 (1947), 22–37.

[7] SURMA S. J.: *Method of natural deduction in equivalential and equivalential-negational propositional calculus*.
Universitas Iagellonica Acta Scientiarum Litterarumque, CCLXV, Schedae Logicae, (6) 1971, 55–68.

[8] SURMA S. J.: *A survey of the results and methods of investigations of the equivalential propositional calculus*.
This volume.

JACEK K. KABZIŃSKI

KOLMOGOROV AND GLIVENKO'S PAPERS ABOUT INTUITIONISTIC LOGIC [1]

Intuitionism as one of the trends in the field of the foundations of mathematics began with the works of L. E. J. Brouwer (1882–1966) and particularly with his work [1] written in 1908. The subject of this work is an attack against the Principle of the excluded middle. Brouwer questioned an application of this principle in reasonings concerned with infinite sets. In this way he eliminated the well-known antinomies of the „naive" set theory of G. Cantor. Of course, it is not the only method to eliminate these antinomies. For example E. Zermelo eliminates them by axiomatic restriction of the notion of set.

Together with the rise of intuitionistic mathematics the problem arose how to reproduce the classical mathematics in it. A. N. Kolmogorov (born 1903) was the first mathematician who dealt with this problem and provided it with a clear logical mathematical form. Here I mean his work *On the Principle of excluded middle* written in 1925 (cf. [8]). These problems were undertaken by V. M. Glivenko (1897–1940) in the works *Sur la logique de M. Brouwer* written in 1928 and *Sur quelques points de la logique de M. Brouwer* written in 1929 (cf. [2] and [3]). It is characteristic that Glivenko does not refer to the paper [8] either in [2] or in [3]. The aim of the discussed paper [8] is to define precisely the range of application of the classical propositional calculus. In this paper three implicational-negational logical calculi formalized by means of substitution and detachment are discussed. The axioms of the first system originate from Hilbert and were presented for the first time in 1922 in the paper [6]. The next system of axioms presented by Kolmogorov is inferentially

[1] This article contains the text of the lecture given by the author on April 25, 1970 to the XVIth Conference of the History of Logic organized in Cracow by the Section of Logic, Polish Academy of Sciences, together with the Department of Logic of the Jagiellonian University.

equivalent to Hilbert's system of axioms. The calculus based upon three axioms is called by Kolmogorov the special logic of judgments. The third system of axioms is poorer than the former one. The calculus based on these axioms is an implicational-negational part of the calculus examined by I. Johansson ten years later 1935 in the paper [7] and later called minimal calculus. Kolmogorov calls it general logic of judgments.

Hilbert's system of axioms is as follows:

(1) $CpCqp$

(2) $CCpCpqCpq$

(3) $CCqrCCpqCpr$

(4) $CCpCqrCqCpr$

(5) $CpCNpq$

(6) $CCpqCCNpqq$

The system of axioms for the special logic of judgments is different from the first one only in the axioms characterizing negation, namely, instead of (5) and (6) it contains the axioms:

(7) $CCpqCCpNqNp$

(8) $CNNpp$

The letters p, q, r denote here propositional variables. In contradistinction to them formulas will be denoted by $\alpha, \beta, \gamma, \ldots$ built up of the variables and the symbols C and N.

Continuing the discussion Kolmogorov gives a detailed proof of the inferential equivalence of the given systems of axioms, namely, the following is proved.

THEOREM 1. *Hilbert's system of axioms* (1)–(6) *is inferentially equivalent to Kolmogorov's system of axioms* (1)–(4), (7), (8) *for the special logic of judgments.*

Kolmogorov's third system of axioms is composed of the axioms (1)–(4) and (7). Kolmogorov discusses in a detailed way the negational axioms of Hilbert's system. In particular he denies axiom (5) intuitive properties. He says: „[...] it is used only in a symbolic presentation of the logic of judgments; therefore it is not affected by Brouwer's critique [...] does not have and cannot have any intuitive foundation [...]". The general logic of judgments of Kolmogorov is so poor that the formulas questioned by him cannot be proved in it. Of course, according to the assumptions of Brouwer's intuitionism, Kolmogorov rejects axiom (6). It is obvious because on the basis of suitably chosen definition of the alternative axiom (6) is equivalent to the Principle of the excluded middle. It appears that for the formulas of some shape in the general logic of judgments axiom (8) can be proved. Namely, let $Cn(X)$ and correspondingly $Cn_M(X)$ denote the smallest sets of propositional formulas containing the set X and the set of all theses of the classical propositional logic and correspondingly the set of all theses of the general logic of judgments whereas each of these sets is closed under the rules of substitution and detachment. Then Kolmogorov theorems have the shape:

THEOREM 2. $CNNNpNp \in Cn_M(\emptyset)$.

THEOREM 3. *If* $CNNaa \in Cn_M(\emptyset)$ *and* $CNN\beta\beta \in Cn_M(\emptyset)$, *then* $CNNCa\beta Ca\beta \in Cn_M(\emptyset)$.

Let a be any implicational-negational formula built up exactly of the variables $p_1, p_2, \ldots, p_n$. Let us denote by the letter (A) the following conditions:

(A) $CNNp_ip_i \in Cn_M(\emptyset)$ for any $1 \leqslant i \leqslant n$.

Kolmogorov really states what may be formulated in the following theorems:

THEOREM 4. *If* (A) *holds, then* $CNNaa \in Cn_M(\emptyset)$.

THEOREM 5. *If* (A) *holds and* $a \in Cn(\emptyset)$, *then* $a \in Cn_M(\emptyset)$.

In the next part of the paper [8] Kolmogorov gives a method of transforming any formula into the formula which fulfills the condition (A). To every formula a taken from the language of any mathematical theory we assign a formula a^* as its interpretation in the following way:

(1) for a variable p we put $p^* = NNp$

(2) for a negational formula Na we put $(Na)^* = NNNa^*$

(3) for a implicational formula $Ca\beta$ we put $(Ca\beta)^* = NNCa^*\beta^*$.

For example:

$$(CCpqCCpNqNp)^* =$$
$$= NNCNNCNNpNNqNNCNNCNNpNNNNNqNNNNNp$$
$$(CNNpp)^* = NNCNNNNNNNpNNp.$$

A formula whose interpretation into the formula fulfilling the condition (A) is a thesis of the general logic of judgments is called a pseudo-true formula. A mathematics whose formulas are interpretations of the mathematical formulas is called by Kolmogorov the mathematics of pseudotruth. It is based on the general logic of judgments. In the mathematics of pseudotruth one can use the implicational-negational counterparts of the Principle of the excluded middle without any danger of the criticisms of the intuitionists. Thus the classical mathematics has been reproduced in the mathematics of pseudotruth.

Let $X = \{X_1, X_2, \ldots, X_k\}$ be a system of axioms of any established mathematical theory. Let us denote by X the set of formulas which are interpretations of the axioms, i.e., the set $\{X_1^*, X_2^*, \ldots, X_k^*\}$. Let us formulate what Kolmogorov had in view in the following.

THEOREM 6. *If* $\alpha \in Cn(X)$, *then* $\alpha^* \in Cn_M(X^*)$.

Kolmogorov observes that the interpretation of any classical theory by means of the above given operation * may be treated as an intuitionistic proof of consistency of this theory.

Let us further observe that the problem of interpretation of the classical theories in the corresponding intuitionistic theories was undertaken in the later years by many authors. For example: K. Gödel in the paper [4] written in 1932, B. Sobociński in 1939 and J. Łukasiewicz in the paper [9] written in 1952, (cf. also [10]).

In the addenda to the paper [8] the author discusses the problem of formalization of the intuitionistic logic of predicates. The starting point is an analysis of the following four theses in the first order predicate calculus:

(9) $CN(x)px(Ex)Npx$

(10) $C(Ex)NpxN(x)px$

(11) $CN(Ex)px(x)Npx$

(12) $C(x)NpxN(Ex)px$

where the symbol (x) is to be read „for every x“ and the symbol (Ex) is to be read „there is x“ and x is an individual variable. The first of given formulas is criticised as being non-intuitive. The subsequently construed allows us to prove only the last three of the given formulas. This calculus is based on the general logic of judgments to which the following axioms are added:

(13) $C(x)CpxqxC(x)px(x)qx$

(14) $C(x)CpqxCp(x)qx$

(15) $C(x)CpxqC(Ex)pxq$

(16) $Cpx(Ex)px$

together with the well-known rule of generalization. Further remarks of Kolmogorov allow us to think that the formula

(17) $C(x)pxpa$

should also be added to the said system.

Like Kolmogorov's paper [8] also Glivenko's papers [2] and [3] do not aim at axiomatizing intuitionistic logic. Axioms are achieved „by the way" somehow in the course of proving some theorems of the intuitionistic logic. Therefore Glivenko does not bother about its independence. In the article [2] written in 1928 he uses the following axioms to prove the introduced theorems:

(7) $CCpqCCpNqNp$

(18) Cpp

(19) $CCpqCCqrCpr$

(20) $CKpqp$

(21) $CKpqq$

(22) $CCrpCCrqCrKpq$

(23) $CpApq$

(24) $CqApq$

(25) $CCprCCqrCApqr$

(26) $CCpNqCqNp$

(27) $CCpqCNqNp$

The set of axioms (7) and (18)–(27) is inferentially equivalent to the set of axioms (1), (2), (7) and (19)–(25) of the Johansson minimal propositional calculus.

The theorems proved by Glivenko can be written as follows:

THEOREM 7. $NNANpp \in Cn_M(\emptyset)$

THEOREM 8. $CNNNpNp \in Cn_M(\emptyset)$

THEOREM 9. $CCANpqNqNq \in Cn_M(\emptyset)$.

In the theorems 7, 8 and 9 the symbol $Cn_M(\emptyset)$ denotes the smallest set containing the axioms (7) and (18)–(27) and closed under the rules of substitution and detachment.

In the article [3] written in 1929 Glivenko uses the set of axioms composed of the formulas (1), (2), (4), (7), (18)–(25) and the formula

(28) $CNpCpq$

to prove the formulated theorems. He proves two important theorems for the calculus with substitution and detachment based on these axioms:

THEOREM 10. *If* $Na \in Cn(\emptyset)$, *then* $Na \in Cn_I(\emptyset)$

THEOREM 11. *If* $a \in Cn(\emptyset)$, *then* $NNa \in Cn_M(\emptyset)$.

In the theorems 10 and 11 the symbol $Cn_I(\emptyset)$ denotes the smallest set containing the axioms (1), (2), (4), (7) and (18)–(25) and (28) and closed under the rules of substitution and detachment.

The last system of axioms is identical with the axioms proposed for the intuitionistic logic by A. Heyting in the paper [5] written in 1930.

References

[1] BROUWER L. E. J.: *De onbetrouwbaarheid der logische principes.* Tijdschrift voor wijsbegeerte, 2(1908), 152–158.

[2] GLIVENKO V. M.: *Sur la logique de M. Brouwer.* Académie Royale de Belgique, Bulletin de la Classe des Sciences, 14 (1928), 225–228.

[3] GLIVENKO V. M.: *Sur quelques points de la logique de M. Brouwer.* Académie Royale de Belgique. Bulletin de la Classe des Sciences, 15(1929), 183–188.

[4] GÖDEL K.: *Zur intuitionistischen Arithmetik und Zahlentheorie.* Ergebnisse eines mathematischen Kolloquiums, 4(1931–1932), 34–38.

[5] HEYTING A.: *Die formalen Regeln der intuitionistischen Logik.* Sitzungsberichte der Preussischen Akademie der Wissenschaften. Physikalisch-mathematische Klasse, 1930, 42–56.

[6] HILBERT D.: *Die logischen Grundlagen der Mathematik.* Mathematische Annalen, 88(1923), 151–165.

[7] JOHANSSON I.: *Der Minimalkalkül, ein reduzierter intuitionistischer Formalismus.* Compositio Mathematica, 4(1936), 119–136.

[8] KOLMOGOROV A. N.: *O principie tertium non datur.* Matematičeskij Sbornik, 32(1924–1925), 646–667.

[9] ŁUKASIEWICZ J.: *O intuicjonistycznym rachunku zdań.* In: *Z zagadnień logiki i filozofii.* Warszawa 1961, 261–274.

[10] PRAWITZ D., MALMNAS P. E.: *A survey of some connections between classical, intuitionistic and minimal logic.* In: *Contributions to mathematical logic.* Amsterdam 1968, 215–229.

STANISŁAW J. SURMA

JAŚKOWSKI'S MATRIX CRITERION FOR THE INTUITIONISTIC PROPOSITIONAL CALCULUS [1]

1. Introduction

In the present paper I shall submit a detailed proof of the Jaśkowski's theorem about the intuitionistic propositional calculus according to which any propositional formula is a theorem of the intuitionistic propositional calculus if and only if it is tautological in each of Jaśkowski's matrices (cf. [2]). Jaśkowski himself expounded only the keynote for the proof of this theorem comprising it within hardly half a page (cf. [2], p. 60). Subsequent publications have noted the difficulties involved in the attempts to reproduce the proof proceeding along the lines of Jaśkowski. For example, G. F. Rose proved the Jaśkowski's theorem by another method „after several unsuccessful attempts to establish the 'Lemma' of [2], p. 60 with the aid of Jaśkowski's rough outline" (cf. [3], p. 8, footnote (16)). Here I shall largely resort to Jaśkowski's main idea as well as to his terminology.

Among the most widely known methods for proving the completeness of propositional calculi it seems that Łukasiewicz's methods in 1929 (cf. [6]) is most comparable to Jaśkowski's methods of proof which I am going to discuss below. This becomes quite understandable when one considers that Jaśkowski was studying under Łukasiewicz's guidance.

---

[1] This paper contains the text of a lecture given by the author on April 24, 1970 to the XVIth Conference of the History of Logic organized in Cracow by the Section of Logic, Polish Academy of Sciences, together with the Department of Logic at the Jagiellonian University.

2. Intuitionistic propositional calculus. Regular formulas.

Small Latin initial letters, with or without indices, are used to denote propositional variables. An infinite sequence of all propositional variables is denoted by *Var*. Propositional connectives of negation, implication, conjunction and disjunction are denoted, respectively, by

$$\neg, \rightarrow, \wedge, \vee .$$

Propositional formulas construed out of variables and connectives are denoted by capital Latin letters. In particular, the negation of formula A and the implication, conjunction and disjunction construed from formulas A and B are denoted, respectively, by

$$\neg A, A \rightarrow B, A \wedge B, A \vee B .$$

Propositional variables and the formulas of the above form, in which A and B are propositional variables are called simple formulas. We are introducing, moreover, the following highly convenient abbreviations:

$$\bigwedge_{i=1}^{i=n} A_i = \bigwedge_1^n A_i = \bigwedge_1^n A = A_1 \wedge A_2 \wedge \ldots \wedge A_n$$

and

$$\bigwedge_1^{n-\{i\}} A_i = A_1 \wedge A_2 \wedge \ldots \wedge A_{i-1} \wedge A_{i+1} \wedge \ldots \wedge A_n .$$

The set of all propositional formulas is denoted by S.

The intuitionistic propositional calculus which was formalized for the first time by A. Heyting (cf. [1]) is founded upon the well-known rules of detachment and substitution as well as upon specially selected axioms made up of connectives and propositional variables. As we know, the rule of substitution may be eliminated from the description of this calculus by replacing the axioms by their schemae, i.e., by the classes of all substitutions of these axioms. These schemae are as follows:

(i_1) $A \rightarrow (B \rightarrow A)$

(i_2) $(A \rightarrow (B \rightarrow C)) \rightarrow ((A \rightarrow B) \rightarrow (A \rightarrow C))$

(i_3) $A \wedge B \rightarrow A$

(i_4) $A \wedge B \rightarrow B$

(i_5) $A \rightarrow (B \rightarrow A \wedge B)$

(i_6) $A \rightarrow A \vee B$

(i_7) $B \to A \vee B$

(i_8) $(A \to C) \to \big((B \to C) \to (A \vee B \to C)\big)$

(i_9) $(A \to B) \to \big((A \to \neg B) \to \neg A\big)$

(i_{10}) $\neg A \to (A \to B)$

In such a substitutionless calculus there holds the classical deduction theorem which remarkably simplifies logical calculi. According to this theorem to prove any implication it is enough to deduce its consequent from its antecedents applying beside the (schemae of) axioms only the rule of detachment. It should be noted that in the intuitionistic propositional calculus (without the rule of substitution) there also holds the so-called weak indirect deduction theorem according to which to prove any implication with the consequent $\neg B$ it is enough to deduce a couple of contradictory formulas from the antecedents of this implication and from the formula B applying the axioms (i_1)–(i_{10}) and the rule of detachment (cf. [5]).

Let us introduce two more convenient abbreviations. Thus, the notation

$$A \leftrightarrow B$$

signifies that formulas A and B are equivalent on the grounds of the intuitionistic propositional calculus with the rule of detachment as the only rule of deduction whereas the notation

$$A \Leftrightarrow B$$

signifies that formulas A and B are equivalent on the grounds of the intuitionistic propositional calculus with the detachment and substitution. Obviously, the condition stating that $A \leftrightarrow B$ entails the condition stating that $A \Leftrightarrow B$ although not always conversely. There holds the easily provable.

LEMMA 1. *In the intuitionistic propositional calculus there hold the following equivalencies:*

(i_{11}) $A \wedge (B \to A) \leftrightarrow A \wedge \big(B \to (C \to A)\big) \leftrightarrow A \wedge \big(B \to (A \vee C)\big) \leftrightarrow$
$A \wedge \big(B \to (C \vee A)\big) \leftrightarrow A \wedge (\neg A \to B) \leftrightarrow A,$

(i_{12}) $A \wedge (A \to B) \leftrightarrow A \wedge \big((C \to A) \to B\big) \leftrightarrow A \wedge B,$

(i_{13}) $A \wedge (B \to \neg A) \leftrightarrow A \wedge \neg B,$

(i_{14}) $A \wedge \big(B \to (A \to C)\big) \leftrightarrow A \wedge \big((A \to B) \to C\big) \leftrightarrow A \wedge (B \to C),$

(i_{15}) $A \wedge (B \vee C) \leftrightarrow (A \wedge B) \vee (A \wedge C),$

(i_{16}) $\big((A \vee B) \to C\big) \leftrightarrow (A \to C) \wedge (B \to C),$

(i_{17}) $(A \wedge B \to C) \leftrightarrow (A \to (B \to C))$,

(i_{18}) $(A \to B \wedge C) \leftrightarrow (A \to B) \wedge (A \to C)$.

LEMMA 2. *If a variable a does not occur in a formula A, then $A \Leftrightarrow ((B \leftrightarrow a) \to A(B/a))$, where the notation $A(B/a)$ denotes the result of replacement of the variable a for the formula B.*

Proof. Let us assume that a does not occur in A and let us note that by means of induction on the construction of A it can be proved that

(1) $A \to ((B \leftrightarrow a) \to A(B/a))$.

On the other hand, substituting in the formula

(2) $(B \leftrightarrow a) \to A(B/a)$

the formula B for the variable a, we obtain

(3) $(B \leftrightarrow B) \to A$.

Hence it follows the formula

(4) A.

From (1), (2) and (4) it results that

(5) $A \Leftrightarrow ((B \leftrightarrow a) \to A(B/a))$

which completes our proof.

LEMMA 3. *For any formula $A \in S$ there exists n such that $A \Leftrightarrow (\bigwedge_1^n (B_i \leftrightarrow b_i) \to a)$, where a and b_i are the mutually distinct variables not occurring in A and where B_i are simple formulas for any $i \leqslant n$.*

The proof of the lemma is inductive. The starting step results from lemma 2, namely,

(I) if A is a simple formula, then $A \Leftrightarrow ((A \leftrightarrow a) \to a)$ where a variable a does not occur in A.

Now, let A contain k connectives, where $k > 1$.
Let us assume that

(II) for some m $A \Leftrightarrow \bigwedge_1^m (B_i \leftrightarrow b_i) \to a$, where for any $i \leqslant m$ variables b_i are mutually distinct and besides do not occur in formula A

and let us suppose that

(III) there exists $j \leqslant m$ such that in formula B_j occur $r+2$ connectives, where $r \geqslant 0$.

Hence

(IV) formula B_j is equal to one of the formulas below: $\neg B$, $B \to C$, $B \vee C$, $B \wedge C$.

First we shall show that

(V) if $B_j = \neg B$, then $\left(\bigwedge_1^m (B_i \leftrightarrow b_i) \to a\right) \Leftrightarrow \left(\bigwedge_1^{m-\{j\}} (B_i \leftrightarrow b_i) \wedge (B \leftrightarrow b) \wedge (\neg b \leftrightarrow b_j) \to a\right)$ where the variable b does not occur in formulas A and $\bigwedge_1^m (B_i \leftrightarrow b_i) \to a$.

Indeed, let us assume

(1) $\bigwedge_1^m (B_i \leftrightarrow b_i) \to a$

and

(2) $\bigwedge_1^{m-\{j\}} (B_i \leftrightarrow b_i) \wedge (B \leftrightarrow b) \wedge (\neg b \leftrightarrow b_j)$.

From (2) it follows

(3) $\neg B \leftrightarrow b_j$.

Hence from (1) and (2) we get

(4) a.

To prove the converse inference let us assume

(5) $\bigwedge_1^{m-\{j\}} (B_i \leftrightarrow b_i) \wedge (B \leftrightarrow b) \wedge (\neg b \leftrightarrow b_j) \to a$

where the variable b does not occur in formulas A and $\bigwedge_1^m (B_i \leftrightarrow b_i) \to a$. Let us also assume

(6) $\bigwedge_1^m (B_i \leftrightarrow b_i)$.

In the case (V) we have put $B_j = \neg B$. Thus, substituting in (5) the formula B for the variable b and omitting the theorem $B \leftrightarrow B$ we obtain

(7) a

which completes the proof of the line (V). Now we shall show that

(VI) if $B_j = B \to C$, then $\bigwedge_1^m (B_i \leftrightarrow b_i) \to a) \Leftrightarrow \left(\bigwedge_1^{m-\{j\}} (B_i \leftrightarrow b_i) \wedge (B \leftrightarrow b) \wedge \wedge (C \leftrightarrow c) \wedge \left((b \to c) \leftrightarrow b_j\right) \to a\right)$ where the mutually distinct variables b, c do not occur in the formulas A and $\bigwedge_1^m (B_i \leftrightarrow b_i) \to a$.

The first inference stated in (VI) follows from the theorem

(1) $\left((B \leftrightarrow b) \wedge (C \leftrightarrow c)\right) \wedge \left((b \to c) \leftrightarrow b_j)\right) \to \left((B \to C) \leftrightarrow b_j\right)$.

The converse inference stated in (VI) results by substitution of the formulas B and C for the variables b and c, respectively. By analogical reasoning one can prove also the lines

(VII) if $B_j = B \vee C$, then $\left(\bigwedge_1^m (B_i \leftrightarrow b_i) \to a\right) \Leftrightarrow \left(\bigwedge_1^{m-\{j\}} (B_i \leftrightarrow b_i) \wedge (B \leftrightarrow b) \wedge \wedge (C \leftrightarrow c) \wedge \left((b \vee c) \leftrightarrow b_j\right) \to a\right)$ where the mutually distinct variables b, c do not occur in the formulas A and $\bigwedge_1^m (B_i \leftrightarrow b_i) \to a$

and

(VIII) if $B_j = B \wedge C$, then $\left(\bigwedge_1^m (B_i \leftrightarrow b_i) \to a\right) \Leftrightarrow \left(\bigwedge_1^{m-\{j\}} (B_i \leftrightarrow b_i) \wedge (B \leftrightarrow b) \wedge \wedge (C \leftrightarrow c) \wedge \left((b \wedge c) \leftrightarrow b_j\right) \to a\right)$ where the mutually distinct variables b, c do not occur in the formulas A and $\bigwedge_1^m (B_i \leftrightarrow b_i) \to a$.

From (IV)–(VIII) it follows that

(IX) for some s $A \Leftrightarrow \left(\bigwedge_1^{m-\{j\}}(B_i \leftrightarrow b_i) \wedge \bigwedge_1^{s}(C_i \leftrightarrow c_i) \rightarrow a\right)$ where for any $i \leqslant s$ in the formula C_i there occur at last $r+1$ connectives and where the mutually distinct variables $c_1, c_2, \ldots, c_s$ do not occur in the formulas A and $\bigwedge_1^{m-\{j\}}(B_i \leftrightarrow b_i)$.

From (I), (II) and (IX) there results, according to the rule of induction, lemma 3.

EXAMPLE. $\neg(a \vee (b \rightarrow c)) \Leftrightarrow \big((a \leftrightarrow c_1) \wedge \big((b \rightarrow c) \leftrightarrow c_2\big) \wedge \big((c_1 \vee c_2) \leftrightarrow c_3\big) \wedge \wedge (\neg c_3 \leftrightarrow c_4) \rightarrow c_4$.

LEMMA 4. *A formula $A \rightarrow \neg B$ is a theorem of the classical propositional calculus if and only if it is a theorem of the intuitionistic propositional calculus.*

The proof of the first inference stated in lemma 2 is inductive on the length of the classical proof of the theorem $A \rightarrow \neg B$. The converse inference stated in that lemma is trivial.

DEFINITION 1. *A formula A is regular if there exists n such that $A = (A_1 \wedge A_2 \wedge \ldots \wedge A_n \rightarrow a)$, where for an arbitrary $i \leqslant n$ either A_i is a variable or A_i is a simple negation or it is an implication built up of at most two connectives* $\neg, \rightarrow, \wedge, \vee$.

From this definition it results that A_i may be equal to one of the formulas below

(1)	a	(7)	$\neg a \rightarrow b$
(2)	$\neg a$	(8)	$(a \rightarrow b) \rightarrow c$
(3)	$a \rightarrow b$	(9)	$(a \vee b) \rightarrow c$
(4)	$a \rightarrow \neg b$	(10)	$a \rightarrow b \wedge c$
(5)	$a \rightarrow (b \rightarrow c)$	(11)	$a \wedge b \rightarrow c$
(6)	$a \rightarrow b \vee c$		

On the grounds of lemma 1 (i_{16}), (i_{17}), (i_{18}) cases (9) and (10) can be reduced to case (3) while case (11) to case (5). This remark will permit to strengthen to some extent the definition of the regular formula and as a result to restrict the class of all regular formulas. Namely, formula A is regular if $A = (A_1 \wedge A_2 \wedge \ldots \wedge A_8 \rightarrow a)$, where there exist $n_1, n_2, \ldots, n_8$ such that $n_1 + n_2 + \ldots + n_8 > 0$ and

$$A_1 = \bigwedge_1^{n_1} a_i \,,$$

$$A_2 = \bigwedge_1^{n_2} \neg a_i\,,$$
$$A_3 = \bigwedge_1^{n_3}(a_i \to b_i)\,,$$
$$A_4 = \bigwedge_1^{n_4}(a_i \to \neg b_i)\,,$$
$$A_5 = \bigwedge_1^{n_5}\big(a_i \to (b_i \to c_i)\big)\,,$$
$$A_6 = \bigwedge_1^{n_6}\big(a_i \to (b_i \vee c_i)\big)\,,$$
$$A_7 = \bigwedge_1^{n_7}(\neg a_i \to b_i)\,,$$
$$A_8 = \bigwedge_1^{n_8}\big((a_i \to b_i) \to c_i\big)\,.$$

Regular formulas will be presented sometimes also in the form

$$A_1 \wedge A_2 \wedge \ldots \wedge A_6 \wedge \bigwedge_1^k (B_i \to b_i) \to a$$

or, still more briefly, in the form

$$A' \wedge \bigwedge_1^k (B_i \to b_i) \to a$$

where formula $A' = A_1 \wedge A_2 \wedge \ldots \wedge A_6$ is defined as above and where $k \geqslant 0$ and for an arbitrary $i \leqslant k$ formula B_i is either simple a negation or a simple implication.

The class of all regular formulas with thus defined index k is denoted by Q_k. Within this class we distinguish the subclass Q_k^* consisting of such formulas that no variable occurring in A_1 is equal to any variable occurring in $A_3 \wedge \ldots \wedge A_6 \wedge \bigwedge_1^k (B_i \to b_i)$.

EXAMPLES.

(1) *formula* $a \wedge (b \vee c) \to d$ *is irregular*

(2) $a \wedge (b \to \neg a) \to b \in Q_0 \setminus Q_0^*$
$a \wedge \neg a \to b \in Q_0^*$

(3) $a \wedge \neg b \wedge (\neg a \to b) \to c \in Q_1 \setminus Q_1^*$
$\neg a \wedge (\neg b \to a) \to b \in Q_1^*$

(4) $a \wedge (\neg a \to b) \wedge \big((a \to b) \to c\big) \to d \in Q_2 \setminus Q_2^*$
$d \wedge (a \to \neg b) \wedge (\neg b \to c) \wedge (\neg a \to b) \to e \in Q_2^*$

LEMMA 5. *If* $A \wedge B \wedge C$ *is a formula such that*

(i) $A = \bigwedge_1^{n_1} a$,

(ii) $B = \bigwedge_1^{n_2} \neg a$,

(iii) $C = \bigwedge_1^{n_3}(a \to D) \wedge \bigwedge_1^{n_4}(E \to a) \wedge \bigwedge_1^{n_5}(a \to b \vee c)$,

where D *is either variable or simple negation or simple implication while* E *is either simple negation or simple implication, then there exist a number* s *and a formula* X *such that*

(iv) $X = (A_1 \wedge B_1 \wedge C_1) \vee \ldots \vee (A_s \wedge B_s \wedge C_s) \leftrightarrow A \wedge B \wedge C$,

(v) *for any* $i \leqslant s$ *the formulas* A_i, B_i, C_i *are built like the formulas* A, B, C, *respectively,*

(vi) *each variable occurring in* A_i *occurs also in the formula* $A \wedge B \wedge C$,

(vii) *each factor in* C_i *having one of the form:* $\rightarrow a \rightarrow b$, $(a \rightarrow b) \rightarrow c$ *occurs as a factor in the formula* C, *and*

(viii) *variables occurring in* A_i *do not occur in* C_i.

Proof. Obviously, for a formula $A \wedge B \wedge C$ fulfilling the conditions (i)–(iii) there exist a number s and a formula X_0 fulfilling the conditions (iv)–(vii) and such that for any $i \leqslant s$ either the number of variables occurring in A_i which do not occur in C_i is equal at least to O or the condition (viii) is fulfilled. Such a formula may be equal, e.g., to $A \wedge B \wedge C$. Now let us assume inductively that there exist a number k and a formula X_n such that

(I) formula $X_n = (A_1 \wedge B_1 \wedge C_1) \vee (A_2 \wedge B_2 \wedge C_2) \vee \ldots \vee (A_k \wedge B_k \wedge C_k)$ fulfills the conditions (iv)–(vii) and for any $i \leqslant k$ either the number of variables occurring in A_i which do not occur in C_i is equal at least to n or the condition (viii) is fulfilled.

We shall define a sequence of formulas $Y_1, Y_2, \ldots, Y_k$. Namely, for any $i \leqslant k$ we shall put

(II) $Y_i = A_i \wedge B_i \wedge C_i$, if for $A_i \wedge B_i \wedge C_i$ the condition (viii) is fulfilled.

In the opposite case the definition of the formulas Y_i is more difficult. Namely, let us suppose that

(1) $A_i \wedge B_i \wedge C_i$ do not fulfill the condition (viii).

This means that there exists a variable a such that

(2) a occurs in A_i as well in C_i.

By virtue of the inductive hypothesis (I) there exist variables $a_1, a_2, \ldots, a_n$ such that

(3) $a_1, a_2, \ldots, a_n$ occur in A_i but they do not occur in C_i.

Let us assume that

(4) F is a conjunction of all these factors in C_i in which the variable a does not occur

and

(5) $G = G_1 \wedge G_2 \wedge \ldots \wedge G_q$ is a conjunction of all the remaining factors in C_i which do not equal to the formula of the form $a \rightarrow b \vee c$ where b, c differ from a.

Let us assume, at last, that

(6) $\bigwedge_1^r(a \to b_i \vee c_i)$ is a conjunction of all the remaining factors in C_i where, of course, variables b_i, c_i differ from a.

Thus we have

(7) $C_i = F \wedge G \wedge \bigwedge_1^r(a \to b_i \vee c_i)$.

From (4) and (5) it follows that for any $p \leqslant q$

(8) G_p equals to one of the formulas below:

(8.1)	$a \to b$	(8.10)	$\neg a \to b$
(8.2)	$b \to a$	(8.11)	$\neg b \to a$
(8.3)	$a \to \neg c$	(8.12)	$(a \to b) \to a$
(8.4)	$c \to \neg a$	(8.13)	$(a \to c) \to b$
(8.5)	$a \to (b \to c)$	(8.14)	$(c \to a) \to b$
(8.6)	$a \to (c \to b)$	(8.15)	$(b \to c) \to a$
(8.7)	$b \to (a \to c)$	(8.16)	$b \to a \vee c$
(8.8)	$c \to (a \to b)$	(8.17)	$b \to c \vee a$
(8.9)	$b \to (c \to a)$		

where b, c differ from a. For any $i \leqslant k$ we shall define inductively a sequence $\{Y_{i,p}\}_{p \leqslant q}$ putting

(9.1) $Y_{i,0} = A_{i,0} \wedge B_{i,0} \wedge C_{i,0} = A_i \wedge B_i \wedge F$

(9.2) $Y_{i,p+1} = Y_{i,p}$, if G_{p+1} equals to the formula of one of the forms (8.2), (8.9), (8.10), (8.11), (8.15), (8.16) or (8.17)

(9.3) $Y_{i,p+1} = A_{i,p} \wedge B_{i,p} \wedge C_{i,p} \wedge b$, if G_{p+1} equals to the formula of one of the forms (8.1), (8.6), (8.8), (8.13) or (8.14)

(9.4) $Y_{i,p+1} = A_{i,p} \wedge B_{i,p} \wedge C_{i,p} \wedge \neg c$, if G_{p+1} equals to the formula of one of the forms (8.3) or (8.4)

(9.5) $Y_{i,p+1} = A_{i,p} \wedge B_{i,p} \wedge C_{i,p} \wedge (b \to c)$, if G_{p+1} equals to the formula of one of the forms (8.5), (8.7) or (8.12).

In the lines (9.1)–(9.5) we have assumed that variables b, c differ from a. On the basis of (2) and (9.1)–(9.5) it is easy to see that

(9) $Y_{i,q} = A_{i,q} \wedge B_{i,q} \wedge C_{i,q} \leftrightarrow A_i \wedge B_i \wedge F \wedge G$.

According to (7) we shall denote by

$$\{H_{1,1}, H_{2,1}, \ldots, H_{r,1}\},$$
$$\{H_{1,2}, H_{2,2}, \ldots, H_{r,2}\},$$
$$\ldots\ldots\ldots\ldots$$
$$\{H_{1,2^r}, H_{2,2^r}, \ldots, H_{r,2^r}\}$$

all the elements of the Cartesian product

$$\{b_1, c_1\}\times\{b_2, c_2\}\times\ldots\times\{b_r, c_r\}.$$

Let us observe that according to (2) a variable a occurs in $Y_{i,q}$. Hence and from (9) and (10) it results that the following equivalences

$$\begin{aligned}(10)\qquad & A_i\wedge B_i\wedge C_i\leftrightarrow\\ & \leftrightarrow Y_{i,q}\wedge(b_1\vee c_1)\wedge\ldots\wedge(b_r\vee c_r)\leftrightarrow\\ & \leftrightarrow Y_{i,q}\wedge\big((H_{1,1}\wedge\ldots\wedge H_{r,1})\vee\ldots\vee(H_{1,2^r}\wedge\ldots\wedge H_{r,2^r})\big)\leftrightarrow\\ & \leftrightarrow(Y_{i,q}\wedge H_{1,1}\wedge\ldots\wedge H_{r,1})\vee\ldots\vee(Y_{i,q}\wedge H_{1,2^r}\wedge\ldots\wedge H_{r,2^r})\leftrightarrow\\ & \leftrightarrow(A_{i,q}\wedge H_{1,1}\wedge\ldots\wedge H_{r,1}\wedge B_{i,q}\wedge C_{i,q})\vee\ldots\\ & \ldots\vee(A_{i,q}\wedge H_{1,2^r}\wedge\ldots\wedge H_{r,2^r}\wedge B_{i,q}\wedge C_{i,q})\end{aligned}$$

hold. According to (10) we shall put

$$\begin{aligned}(\mathrm{III})\qquad & Y_i=(A_{i,q}\wedge H_{1,1}\wedge\ldots\wedge H_{r,1}\wedge B_{i,q}\wedge C_{i,q})\vee\ldots\\ & \ldots\vee(A_{i,q}\wedge H_{1,2^r}\wedge\ldots\wedge H_{r,2^r}\wedge B_{i,q}\wedge C_{i,q}),\end{aligned}$$

if for $A_i\wedge B_i\wedge C_i$ the condition (viii) is not fulfilled.

By means of the formulas $Y_1, Y_2, \ldots, Y_k$ defined in (II) and (III) we shall define a formula X_{n+1} putting

(IV) $X_{n+1}=Y_1\vee Y_2\vee\ldots\vee Y_k$.

Let us observe on the basis of (II) and (III) that for any $i\leqslant k$

(V) $Y_i\leftrightarrow A_i\wedge B_i\wedge C_i$.

Hence, from the inductive hypothesis (I) and from (II.3) it results that for some m

(VI) formula $A\wedge B\wedge C\leftrightarrow X_{n+1}=Y_1\vee Y_2\vee\ldots\vee Y_k=$
$=(A'_1\wedge B'_1\wedge C'_1)\vee\ldots\vee(A'_m\wedge B'_m\wedge C'_m)$
fulfills the conditions (iv)–(vii) and besides for any $j\leqslant m$ either the number of variables occurring in A'_j which do not occur in C'_j is equal at least to $n+1$ or the condition (viii) is fulfilled.

Thus we have proved that for any formula $A\wedge B\wedge C$ fulfilling the conditions (i)–(iii) and for any number n there exist a number k and formula

$$X_n=(A_{1,n}\wedge B_{1,n}\wedge C_{1,n})\vee\ldots\vee(A_{k,n}\wedge B_{k,n}\vee C_{k,n})$$

such that the conditions (iv)–(vii) are fulfilled and for any $j \leqslant k$ either the number of the variables occurring in $A_{j,n}$ which do not occur in $C_{j,n}$ is equal to at least n or the condition (viii) is fulfilled. By virtue of the condition (vi) the number of the variables occurring in $A_{j,n}$ cannot be greater than s where s is the number of all the variables occurring in $A \wedge B \wedge C$. Thus the formula

$$X_{s+1} = (A_{1,s+1} \wedge B_{1,s+1} \wedge C_{1,s+1}) \vee \ldots \vee (A_{t,s+1} \wedge B_{t,s+1} \wedge C_{t,s+1})$$

fulfills all the conditions (vi)–(viii) because for any $j \leqslant t$ the number of the variables occurring in $A_{j,s+1}$ which do not occur in $C_{j,s+1}$ must be smaller than $s+1$. This completes the proof of lemma 5[2].

LEMMA 6. *For any* $A \in Q_k$ *there exist* n *and* $A_1, A_2, \ldots, A_n$ *such that* $A \leftrightarrow A_1 \wedge A_2 \wedge \ldots \wedge A_n$ *where for any* $i \leqslant n$ *there exists* $n_i \leqslant k$ *such that* $A_i \in Q^*_{n_i}$.

Proof. Let us assume that

(1) $A \in Q_k$.

Hence and from the definition of the class Q_k

(2) $A = \bigwedge_1^{n_1} a \wedge \bigwedge_1^{n_2} \neg a \wedge \bigwedge_1^{n_3}(a \to B) \wedge \bigwedge_1^{k}(C \to a) \wedge \bigwedge_1^{n_5}(a \to b \vee c) \to d$

while hence and from lemma 5 it results that there exist n and A_1, $A_2, \ldots, A_n$ such that for any $i \leqslant n$

(3) $A_i = \bigwedge_1^{n_{i,1}} a \wedge A' \wedge \bigwedge_1^{n_i}(C \to a)$

(4) $n_i \leqslant k$

(5) variables occurring in $\bigwedge_1^{n_{i,1}} a$ do not occur in $A' \wedge \bigwedge_1^{n_i}(C \to a)$

and

(6) $A \leftrightarrow \big((A_1 \vee \ldots \vee A_n) \to d\big)$.

Finally from (6) and lemma 1 (i_{16}) it follows that

(7) $A \leftrightarrow (A_1 \to d) \wedge (A_2 \to d) \wedge \ldots \wedge (A_n \to d)$

which completes the proof.

EXAMPLE. Let $A = a \wedge b \wedge (c \to d) \wedge (a \to c \vee e) \wedge (b \to a_1 \vee d) \to b_1$. *Let us observe that* $A \in Q_0 \setminus Q_0^*$. *Applying lemma* 1 ($i_{12}$), ($i_{15}$) to A *we obtain*

$$\big(a \wedge b \wedge c \wedge a_1 \wedge (c \to d) \to b_1\big) \wedge \big(a \wedge b \wedge c \wedge d \wedge (c \to d) \to b_1\big) \wedge$$
$$\wedge \big(a \wedge b \wedge e \wedge a_1 \wedge (c \to d) \to b_1\big) \wedge \big(a \wedge b \wedge d \wedge e \wedge (c \to d) \to b_1\big).$$

---

[2] I am very obliged to dr A. Wroński for his suggestion of the method of proving the lemma 5.

Applying to this formula lemma $1\,(i_{11}),\ (i_{12})$ *we obtain that*

$$A \leftrightarrow (a\wedge b\wedge c\wedge d\wedge a_1 \to b_1)\wedge(a\wedge b\wedge c\wedge d\to b_1)\wedge$$
$$\wedge\big(a\wedge b\wedge e\wedge a_1\wedge(c\to d)\to b_1\big)\wedge(a\wedge b\wedge d\wedge e\to b_1)$$

where all the formulas $a\wedge b\wedge c\wedge d\wedge a_1\to b_1$, $a\wedge b\wedge c\wedge d\to b_1$, $a\wedge b\wedge e\wedge a_1(c\to d)\to$ $\to b_1$, $a\wedge b\wedge d\wedge e\to b_1$ *belong to* Q_0^*.

LEMMA 7. *For any* $A\in S$ *there exist* k *and* $B\in Q_k^*$ *such that* $A\Leftrightarrow B$.

Proof. Let $A\in S$. According to lemma 3 there exists n such that

(1) $A\Leftrightarrow\big(\bigwedge_1^n(B_i\leftrightarrow b_i)\to a\big)$

where for any $i\leqslant n$

(2) B_i is a simple formula.

From (2) and from the definition of the abbreviation $\leftrightarrow$ it results that the formula $\bigwedge_1^n(B_i\leftrightarrow b_i)\to a$ can contain, i. a., factors of the forms below

(i) $(c_i\wedge d_i)\to b_i$

(ii) $b_i\to(c_i\wedge d_i)$

(iii) $(c_i\vee d_i)\to b_i$.

Using lemma 1 (i_{16}), (i_{17}), (i_{18}) we replace formulas (i)–(iii) by the formulas below, respectively,

(iv) $c_i\to(d_i\to b_i)$

(v) $(b_i\to c_i)\wedge(b_i\to d_i)$

(vi) $(c_i\to b_i)\wedge(d_i\to b_i)$.

As a result we obtain according to the definition of Q_k^* the formula B such that

(3) $B = A_3\wedge\ldots\wedge A_6\wedge\ \bigwedge_1^k(B_i\to b_i)\to a$

(4) $B\in Q_k^*$

and

(5) $k\leqslant n$

which completes the proof.

EXAMPLE. $\neg\big(a\vee(b\to c)\Leftrightarrow B$, *where*
$B = (a\leftrightarrow d)\wedge(d\to a_1)\wedge(e\to a_1)\wedge(b_1\leftrightarrow\neg a_1)\wedge\big(e\to(b\to c)\big)\wedge(a_1\to d\vee e)$
where $B\in Q_2^*$.

DEFINITION 2. For any $k>0$, $A\in Q_k$ and $i\leqslant k$ we put

$$[A]_i = [A_1\wedge\ldots\wedge A_6\wedge\ \bigwedge_1^k(B_i\to b_i)\to a]_i =$$
$$= A_1\wedge\ldots\wedge A_6\wedge\ \bigwedge_1^k(B_i\to b_i)\to B_i\ .$$

EXAMPLES.

(1) $[a\wedge(\neg a\to b)\to b]_1 = a\wedge(\neg a\to b)\to\neg a$,

(2) $[a\wedge((a\to b)\to c)\to b]_1 = a\wedge((a\to b)\to c)\to(a\to b)$,

(3) $[a\wedge(\neg a\to b)\wedge((a\to b)\to c)\to d]_1 =$
$= a\wedge(\neg a\to b)\wedge((a\to b)\to c)\to\neg a$,
$[a\wedge(\neg a\to b)\wedge((a\to b)\to c)\to d]_2 =$
$= a\wedge(\neg a\to b)\wedge((a\to b)\to c)\to(a\to b)$.

3. Jaśkowski's matrices

DEFINITION 3. *Sequence*

$$M = \{A_M, b_M, \neg_M, \to_M, \wedge_M, \vee_M\}$$

is called a matrix with the designated element b_M *if*

(i) A_M *is a set,*

(ii) $b_M \notin A_M$,

(iii) $\neg_M$ *is a one-argument operation while* $\to_M$, $\wedge_M$ *and* $\vee_M$ *are binary operations in* $A_M+\{b_M\}$.

In the sequel we shall often write $a \in M$ instead of $a \in A_M+\{b_M\}$.

EXAMPLE. $L_0 = \{\emptyset, 1, \neg_0, \to_0, \wedge_0, \vee_0\}$ *is a trivial matrix where* $\emptyset$ *denotes the empty set.*

DEFINITION 4. *Let* M *be a matrix. Then*

(i) *operation* $\neg_M$ *is normal in* M *if* $\neg_M b_M \neq b_M$,

(ii) *operation* $\to_M$ *is normal in* M *if for any* $a \in M$ *from the fact that* $(b_M \to_M a) = b_M$ *it follows that* $a = b_M$,

(iii) *operation* $\wedge_M$ *is normal in* M *if for any* $a, c \in M$ *from the fact that* $(a\wedge_M c) = b_M$ *it follows that* $a = b_M = c$,

(iv) *operation* $\vee_M$ *is normal in* M *if for any* $a, c \in M$ *from the fact that* $(a\vee_M c) = b_M$ *it follows that* $a = b_M$ *or* $c = b_M$.

Matrix M is said to be normal provided that each of its operations is normal.

EXAMPLE. *Matrix* L_0 *from the above example is normal.*

DEFINITION 5. *By the symbol* $E(M)$ *we shall denote the content of the matrix* M *and define it as the set consisting of all* $A \in S$ *such that for any mapping* $h_0: Var \to M$ *and for any homomorphic extension* $h: S \to M$ *of the mapping* h_0, i.e., *for any mapping* $h: S \to M$ *such that for any* $a \in Var$ *and for any* $B, C \in S$:

(i) $h(a) = h_0(a)$,

(ii) $h(\neg B) = \neg_M h(B)$,

(iii) $h(B \to C) = h(B) \to_M h(C)$,

(iv) $h(B \wedge C) = h(B) \wedge_M h(C)$,

(v) $h(B \vee C) = h(B) \vee_M h(C)$,

the equality $h(A) = b_M$ *holds.*

In the sequel we denote both above mappings, h_0 and h, by one and the same letter.

EXAMPLE. $E(L_0) = S$.

DEFINITION 6. *By* a *we denote the function on* M *defined thus:*

(i) $a(b_M) \notin M$,

(ii) *if* $a \neq b_M$, *then* $a(a) = a$ *for any* $a \in M$.

DEFINITION 7. *Let* M *be a matrix. The sequence*

$$\Gamma(M) = \{A_{\Gamma(M)}, b_{\Gamma(M)}, \neg_{\Gamma(M)}, \to_{\Gamma(M)}, \wedge_{\Gamma(M)}, \vee_{\Gamma(M)}\}$$

is the matrix such that

(i) $A_{\Gamma(M)} = A_M + \{a(b_M)\}$,

(ii) $b_{\Gamma(M)} = b_M$,

(iii) *operations in* $\Gamma(M)$ *are defined thus:*

	$\to_{\Gamma(M)}$	$\wedge_{\Gamma(M)}$	$\vee_{\Gamma(M)}$	$\neg_{\Gamma(M)}$
$b_{\Gamma(M)}, b_{\Gamma(M)}$	$b_M \to_M b_M$	$b_M \wedge_M b_M$	$b_M \vee_M b_M$	$a(\neg_M b_M)$
$a(a), b_{\Gamma(M)}$	$a \to_M b_M$	$a(a \wedge_M b_M)$	$a \vee_M b_M$	$\neg_M a$
$b_{\Gamma(M)}, a(c)$	$a(b_M \to_M c)$	$a(b_M \wedge_M c)$	$b_M \vee_M c$	
$a(a), a(c)$	$a \to_M c$	$a(a \wedge_M c)$	$a(a \vee_M c)$	

for any $a, c \in M$.

EXAMPLE. *Matrix*

$$L_1 = \Gamma(L_0) = \{\{2\}, 1, \neg_1, \to_1, \wedge_1, \vee_1\}$$

is defined by the table below where $a(1) = 2$

	$\to_1$	$\wedge_1$	$\vee_1$	$\neg_1$
$1,\ 1$	1	1	1	2
$a(1),\ 1$	1	2	1	1
$1,\ a(1)$	2	2	1	
$a(1),\ a(1)$	1	2	2	

As we see, L_1 *is a two-element Boolean algebra. Thence this is a normal matrix while* $E(L_1)$ *is equal, in conformity with the well-known Post's theorem about the completeness of the classical propositional calculus, to the set of all theorems of the classical propositional calculus.*

LEMMA 8. *If* M *is a normal matrix, then*

(i) $\neg_{\Gamma(M)} a = \neg_M a$,

(ii) $(a \to_{\Gamma(M)} c) = (a \to_M c)$,

(iii) $(a \wedge_{\Gamma(M)} c) = (a \wedge_M c)$,

(iv) $(a \vee_{\Gamma(M)} c) = (a \vee_M c)$,

for any $a, c \in M$.

Proof. Let us observe, first, that

(1) if $a = b_{\Gamma(M)}$, then $\neg_{\Gamma(M)} a = \neg_M a$.

Indeed, from the fact that

(1.1) $a = b_{\Gamma(M)}$

and from definition 4 it results that

(1.2) $\neg_M a \neq a$.

Hence and from definitions 4, 6 and 7 it follows that

(1.3) $\neg_{\Gamma(M)} a = a(\neg_M a) = \neg_M a$.

Let us, further, observe that

(2) if $a \neq b_{\Gamma(M)}$, then $\neg_{\Gamma(M)} a = \neg_M a$.

Indeed, from the fact that

(2.1) $a \neq b_{\Gamma(M)}$

and from definitions 6 and 7 it results that

(2.2) $a(a) = a$.

Hence and from definition 7

(2.3) $\neg_{\Gamma(M)} a = \neg_{\Gamma(M)} a(a) = \neg_M a$.

In view of (1) and (2)

(I) $\neg_{\Gamma(M)} a = \neg_M a$ for any $a \in M$.

Now we shall prove that

(II) $(a \rightarrow_{\Gamma(M)} c) = (a \rightarrow_M c)$ for any $a, c \in M$.

It is obvious that the disjunction

(1) either $a = b_{\Gamma(M)} = c$ or $a = b_{\Gamma(M)} \neq c$
or $a \neq b_{\Gamma(M)} = c$ or $a \neq b_{\Gamma(M)} \neq c$

holds. From definition 7 it results immediately that

(2) if $a = b_{\Gamma(M)} = c$, then $(a \rightarrow_{\Gamma(M)} c) = (a \rightarrow_M c)$.

Now let us assume that

(3.1) $a = b_{\Gamma(M)} \neq c$.

Then according to definition 6

(3.2) $a(c) = c$.

Hence and from the normality of matrix M

3.3) $(a \rightarrow_M c) \neq b_M$.

Hence

(3.4) $a(a \rightarrow_M c) = (a \rightarrow_M c)$.

(Hence and from definition 7

(3.5) $(a \rightarrow_{\Gamma(M)} c) = \big(a \rightarrow_{\Gamma(M)} a(c)\big) = a(a \rightarrow_M c) = (a \rightarrow_M c)$.

Thus

(3) if $a = b_{\Gamma(M)} \neq c$, then $(a \rightarrow_{\Gamma(M)} c) = (a \rightarrow_M c)$.

From definitions 6 and 7 it follows immediately that

(4) if $a \neq b_{\Gamma(M)} = c$ or $a \neq b_{\Gamma(M)} \neq c$,
then $(a \rightarrow_{\Gamma(M)} c) = (a \rightarrow_M c)$

which completes the proof of the line (II). By analogous arguing one can also prove that

(III) $(a \wedge_{\Gamma(M)} c) = (a \wedge_M c)$ for any $a, c \in M$

and

(IV) $(a \vee_{\Gamma(M)} c) = (a \vee_M c)$ for any $a, c \in M$

which completes the proof of the lemma.

LEMMA 9. *If operation* $\neg_M$ *is normal in matrix* M, *then operation* $\neg_{\Gamma(M)}$ *is normal in matrix* $\Gamma(M)$.

Proof. From the normality of operation $\neg_M$ in matrix M it follows that

(1) $\neg_M b_M \neq b_M$.

Hence, from lemma 8 and from definition 7

(2) $\neg_{\Gamma(M)} b_{\Gamma(M)} = \neg_M b_M$.

Hence and from (1) it follows according to definition 7 that

(3) $\quad \neg_{\Gamma(M)} b_{\Gamma(M)} \neq b_{\Gamma(M)}$

which means that $\neg_{\Gamma(M)}$ is normal in $\Gamma(M)$.

LEMMA 10. *Operations* $\rightarrow_{\Gamma(M)}, \wedge_{\Gamma(M)}, \vee_{\Gamma(M)}$ *are normal in every matrix* $\Gamma(M)$.

Proof. Let us assume that for any $a \in M$

(1) $\quad a \neq b_{\Gamma(M)}$.

Hence and from definition 6

(2) $\quad \alpha(a) = a$

and hence and from definition 7 and 6

(3) $\quad (b_{\Gamma(M)} \rightarrow_{\Gamma(M)} a) = \alpha(b_M \rightarrow_M a) \neq b_{\Gamma(M)}$.

Thus

(I) $\quad$ operation $\rightarrow_{\Gamma(M)}$ is normal in $\Gamma(M)$.

Now let

(1) $\quad a \neq b_{\Gamma(M)}$ or $c \neq b_{\Gamma(M)}$.

Then according to definitions 6 and 7

(2) $\quad \alpha(a) = a$ or $\alpha(c) = c$

and in consequence

(3) $\quad (a \wedge_{\Gamma(M)} c) = \alpha(a \wedge_M c) \neq b_{\Gamma(M)}$.

Thus

(II) $\quad$ operation $\wedge_{\Gamma(M)}$ is normal in $\Gamma(M)$.

Let, finally,

(1) $\quad a \neq b_{\Gamma(M)} \neq c$.

Hence and from definitions 6 and 7 it follows that

(2) $\quad (a \vee_{\Gamma(M)} c) = \alpha(a \vee_M c) \neq b_{\Gamma(M)}$.

Thus also

(III) $\quad$ operation $\vee_{\Gamma(M)}$ is normal in $\Gamma(M)$

which completes the proof.

DEFINITION 8. *Sequence*

$$P_{i \in I} M_i = \{A_{P_{i \in I}}, b_{P_{i \in I}}, \neg_{P_{i \in I}}, \rightarrow_{P_{i \in I}}, \wedge_{P_{i \in I}}, \vee_{P_{i \in I}}\}$$

is called the Cartesian product of a sequence of matrices $\{M_i\}_{i \in I}$, *where* I *is a finite set of indices if*

(i) $\quad A_{P_{i \in I}}$ *is the Cartesian product of a sequence of sets* $\{A_{M_i}\}_{i \in I}$,

(ii) $\quad b_{P_{i \in I}} = \{b_{M_i}\}_{i \in I}$,

(iii) $\rightharpoondown_{P_{i\in I}}\{a_i\}_{i\in I} = \{\rightharpoondown_{M_i}\}_{i\in I}$,

(iv) $\{a_i\}_{i\in I} \rightarrow_{P_{i\in I}} \{c_i\}_{i\in I} = \{a_i \rightarrow_{M_i} c_i\}_{i\in I}$,

(v) $\{a_i\}_{i\in I} \wedge_{P_{i\in I}} \{c_i\}_{i\in I} = \{a_i \wedge_{M_i} c_i\}_{i\in I}$,

(vi) $\{a_i\}_{i\in I} \vee_{P_{i\in I}} \{c_i\}_{i\in I} = \{a_i \vee_{M_i} c_i\}_{i\in I}$,

for any $a_i, c_i \in M_i$.

The product of two matrices M and N will be denoted by $M \times N$ and, in particular, product $M \times M$ will be denoted by M^2.

LEMMA 11. *Operations* $\rightharpoondown_{P_{i\in I}}$, $\rightarrow_{P_{i\in I}}$ *and* $\wedge_{P_{i\in I}}$ *in a product* $P_{i\in I} M_i$ *of a sequence of normal matrices* $\{M_i\}_{i\in I}$ *are normal operations while operation* $\vee_{P_{i\in I}}$ *is not normal in this product.*

Proof. Let us assume that for any $i \in I$ operation $\rightharpoondown_M$ is normal in M_i. From definitions 4 and 8 it follows then that $\rightharpoondown_{P_{i\in I}} b_{P_{i\in I}} = \rightharpoondown_{P_{i\in I}}\{b_{M_i}\}_{i\in I} = \{\rightharpoondown_{M_i} b_{M_i}\}_{i\in I} \neq \{b_{M_i}\}_{i\in I} = b_{P_{i\in I}}$. Thus

(I) operation $\rightharpoondown_{P_{i\in I}}$ is normal in $P_{i\in I} M_i$.

Let us, further, assume that for any $i \in I$ operation $\rightarrow_{M_i}$ is normal in M_i and let

(1) $(b_{P_{i\in I}} \rightarrow_{P_{i\in I}} \{a_i\}_{i\in I}) = b_{P_{i\in I}}$.

From definition 8 it follows that

(2) $(b_{P_{i\in I}} \rightarrow_{P_{i\in I}} \{a_i\}_{i\in I}) = \{b_{M_i}\}_{i\in I} \rightarrow_{P_{i\in I}} \{a_i\}_{i\in I} = \{b_{M_i} \rightarrow_{M_i} a\}_{i\in I}$.

From (1) and (2) it follows that

(3) $\{b_{M_i} \rightarrow_{M_i} a_i\} = \{b_{M_i}\}_{i\in I}$.

Hence for any $i \in I$

(4) $(b_{M_i} \rightarrow_{M_i} a_i) = b_{M_i}$.

Hence and from the fact that M_i is normal

(5) $a_i = b_{M_i}$

and hence and from definition 8

(6) $\{a_i\}_{i\in I} = b_{P_{i\in I}}$.

Thus

(II) operation $\rightarrow_{P_{i\in I}}$ is normal in $P_{i\in I} M_i$.

By analogous arguing from the fact that operation $\wedge_{M_i}$ is normal (for any $i \in I$) it follows that

(III) operation $\wedge_{P_{i\in I}}$ is normal in $P_{i\in I} M_i$.

To complete the proof let us observe that the fact that from the normality of operations $\vee_{M_i}$ in M_i (for any $i \in I$) the normality of operation $\vee_{P_{i\in I}}$ in $P_{i\in I} M_i$ does not follow, is entailed by the non-distributivity of general quantifier with respect to disjunction.

LEMMA 12. *For an arbitrary sequence of matrices* $\{M_i\}_{i \in I}$ *the equality*

$$E(P_{i \in I} M_i) = P_{i \in I} E(M_i)$$

holds.

Proof. Let $A \in S$. Let us assume that

(1) $A \in E(P_{i \in I} M_i)$.

Then for any homomorphism $h_{P_{i \in I}}: S \to P_{i \in I} M_i$ satisfying the conditions of definition 5 we have that

(2) $h_{P_{i \in I}}(A) = b_{P_{i \in I}}$.

As we know, this homomorphism determines the sequence $\{h_{M_i}\}_{i \in I}$ of homomorphisms $h_{M_i}: S \to M_i$ satisfying the conditions of definition 5 and such that

(3) $h_{M_i}(A) = b_{M_i}$.

From (3) it follows that for any $i \in I$

(4) $A \in E(M_i)$

hence

(5) $A \in P_{i \in I} E(M_i)$.

To prove the converse inference let us observe that from (5) it results (4) while from (4) (3) results. Let us define the mapping $h_{P_{i \in I}}: Var \to P_{i \in I} M_i$ by means of the equalities

(6) $h_{P_{i \in I}}(a) = \{h_{M_i}(a)\}_{i \in I}$

for any $a \in Var$. By induction on the construction of A (for this matter compare a quite analogous proof of lemma 14 below) one can prove then that

(7) $h_{P_{i \in I}}(A) = \{h_{M_i}(A)\}_{i \in I}$.

From (3) and (7) we obtain by means of definition 5 that

(8) $A \in E(P_{i \in I} M_i)$

which completes the proof.

DEFINITION 9. *The sequence of matrices* $\{J_k\}_{k \geqslant 0}$ *defined by means of the equalities*

(i) $J_0 = L_1$,

(ii) $J_{k+1} = \Gamma\big((J_k)^{k+1}\big)$,

is called the sequence of Jaśkowski's matrices.

Let us notice that Jaśkowski's matrix J_k has $\Big(\ldots\Big(\Big(\big((2^1+1)^2+1\big)^3+ +1\Big)^4+1\Big)^5\ldots\Big)^k+1$ elements. In particular, matrices J_0, J_1, J_2, J_3, J_4

have 2, 3, 10, 1001, and 1004006004002 elements, respectively. Thus, the number of elements in Jaśkowski's matrix is rapidly increasing with the increase of k.

For the sake of simplicity we will agree to note k-th Jaśkowski's matrix thus

$$J_k = \{A_k, b_k, \rightharpoondown_k, \rightarrow_k, \wedge_k, \vee_k\}.$$

Further, we shall agree that the $(k+1)$-multiple Cartesian power of the matrix J_k, i.e., matrix $(J_k)^{k+1}$ will be briefly noted as P_{k+1}.

LEMMA 13. *For an arbitrary* $k \geqslant 0$ *the matrix* J_k *is normal.*

Proof. By induction on k we shall first prove, that

(I) operation $\rightharpoondown_k$ is normal in J_k for any $k \geqslant 0$. Let us observe that J_0 is normal (compare one of the examples above). From this it results that

(1) operation $\rightharpoondown_0$ is normal in J_0.

Let us inductively assume that

(2) operation $\rightharpoondown_k$ is normal in J_k.

From (2) by means of lemma 11 it follows that

(3) operation $\rightharpoondown_{P_{k+1}}$ is normal in P_{k+1}

where, as we noted above, $P_{k+1} = (J_k)^{k+1}$. From (3) by means of lemma 9 and definition 9 it follows that

(4) operation $\rightharpoondown_{k+1}$ is normal in J_{k+1}.

We shall further prove that

(II) operation $\rightarrow_k$ is normal in J_k for any $k \geqslant 0$.

Indeed, the matrix J_0 is normal one. Thus

(1) operation $\rightarrow_0$ is normal in J_0.

By means of lemma 10 and definition 9 we infer that

(2) operation $\rightarrow_{k+1}$ is normal in J_{k+1}.

By analogous arguing we obtain that

(III) operation $\wedge_k$ is normal in J_k for any $k \geqslant 0$

and

(IV) operation $\vee_k$ is normal in J_k for any $k \geqslant 0$ which completes the proof.

LEMMA 14. *If* $h_k\colon Var \rightarrow J_k$, *and if* $h_{k+1}\colon Var \rightarrow J_{k+1}$ *is defined by equality*

$$h_{k+1}(a) = \{(h_k(a))_i\}_{i \leqslant k+1}$$

for any $a \in Var$, *then the equality*

$$h_{k+1}(A) = \{(h_k(A))_i\}_{i \leqslant k+1}$$

holds for any $A \in S$.

Proof. We shall prove the lemma by induction on the construction of the formula A. From the hypothesis of the lemma we have that

(I) if $A = a$, then $h_{k+1}(A) = \{(h_k(A))_i\}_{i \leqslant k+1}$.

Let us inductively assume that

(II) if $A = B_1$ or if $A = B_2$, then
$h_{k+1}(A) = \{(h_k(A))_i\}_{i \leqslant k+1}$

under this assumption we shall prove that

(III) if $A = \neg B_1$, then $h_{k+1}(A) = \{(h_k(A))_i\}_{i \leqslant k+1}$.

Let us observe that according to lemma 13 and lemma 11

(1) operation $\neg_{P_{k+1}}$ is normal in P_{k+1}.

From definitions 5 and 9, from (1), from lemma 8, from the inductive hypothesis (II) and, at last, from definitions 8 and 5 we infer the equalities below

$$\begin{aligned}(2)\quad h_{k+1}(A) &= h_{k+1}(\neg B_1) = \neg_{k+1} h_{k+1}(B_1) = \neg_{\Gamma(P_{k+1})} h_{k+1}(B_1) = \\ &= \neg_{P_{k+1}} h_{k+1}(B_1) = \neg_{P_{k+1}}\{(h_k(B_1))_i\}_{i \leqslant k+1} = \\ &= \{(\neg_k h_k(B_1))_i\}_{i \leqslant k+1} = \{(h_k(\neg B_1))_i\}_{i \leqslant k+1} = \\ &= \{(h_k(A))_i\}_{i \leqslant k+1}\end{aligned}$$

which completes the proof of the line (III). Let us further notice that from lemmas 13, 11 and 8 and from (II) there follow the equalities below

$$\begin{aligned}h_{k+1}(A) &= h_{k+1}(B_1 \to B_2) = h_{k+1}(B_1) \to_{k+1} h_{k+1}(B_2) = \\ &= h_{k+1}(B_1) \to_{\Gamma(P_{k+1})} h_{k+1}(B_2) = h_{k+1}(B_1) \to_{P_{k+1}} h_{k+1}(B_2) = \\ &= \{(h_k(B_1))_i\}_{i \leqslant k+1} \to_{P_{k+1}} \{(h_k(B_2))_i\}_{i \leqslant k+1} = \\ &= \{(h_k(B_1) \to_k h_k(B_2))_i\}_{i \leqslant k+1} = \\ &= \{(h_k(B_1 \to B_2))_i\}_{i \leqslant k+1} = \{(h_k(A))_i\}_{i \leqslant k+1}\end{aligned}$$

which proves that

(IV) if $A = B_1 \to B_2$, then $h_{k+1}(A) = \{(h_k(A))_i\}_{i \leqslant k+1}$.

By analogous arguing one can prove that

(V) if $A = B_1 \wedge B_2$, then $h_{k+1}(A) = \{(h_k(A))_i\}_{i \leqslant k+1}$.

At last we shall prove that

(VI) if $A = B_1 \vee B_2$, then $h_{k+1}(A) = \{(h_k(A))_i\}_{i \leqslant k+1}$.

From the hypothesis of the lemma and from definition 5 we infer that

(1) $h_k(B_1),\ h_k(B_2) \in J_k$.

Hence by means of lemma 13 and definition 4

(2) if $h_k(B_1) \vee_k h_k(B_2) = b_k$, then $h_k(B_1) = b_k$, or $h_k(B_2) = b_k$.

Let us assume that

(3.1) $h_{k+1}(B_1) \vee_{P_{k+1}} h_{k+1}(B_2) = b_{P_{k+1}}$.

Hence and from the inductive hypothesis (II)

(3.2) $\{(h_k(B_1))_i\}_{i\leqslant k+1} \vee_{P_{k+1}} \{(h_k(B_2))_i\}_{i\leqslant k+1} = b_{P_{k+1}}$

and hence by means of definition 8

(3.3) $\{(h_k(B_1) \vee_k h_k(B_2))_i\}_{i\leqslant k+1} = b_{P_{k+1}}$.

From (3.3) and (2) it follows according to definition 8 that

(3.4) $\{(h_k(B_1))_i\}_{i\leqslant k+1} = b_{P_{k+1}}$, or $\{(h_k(B_2))_i\}_{i\leqslant k+1} = b_{P_{k+1}}$

hence and from the inductive hypothesis (II)

(3.5) $h_{k+1}(B_1) = b_{P_{k+1}}$, or $h_{k+1}(B_2) = b_{P_{k+1}}$.

Thus

(3) operation $\vee_{P_{k+1}}$ is normal in P_{k+1}.

Hence and from lemma 8 and the inductive hypothesis (II) there follow the equalities below

$$\begin{aligned} h_{k+1}(A) &= h_{k+1}(B_1 \vee B_2) = h_{k+1}(B_1) \vee_{k+1} h_{k+1}(B_2) = \\ &= h_{k+1}(B_1) \vee_{\Gamma(P_{k+1})} h_{k+1}(B_2) = h_{k+1}(B_1) \vee_{P_{k+1}} h_{k+1}(B_2) = \\ &= \{(h_k(B_1))_i\}_{i\leqslant k+1} \vee_{P_{k+1}} \{(h_k(B_2))_i\}_{i\leqslant k+1} = \\ &= \{(h_k(B_1) \vee_k h_k(B_2))_i\}_{i\leqslant k+1} = \{(h_k(B_1 \vee B_2))_i\}_{i\leqslant k+1} = \\ &= \{(h_k(A))_i\}_{i\leqslant k+1} \end{aligned}$$

which completes the proof.

Lemma 15. *If A is a theorem of the intuitionistic propositional calculus, then $A \in E(J_k)$ for any $k \geqslant 0$.*

Proof. By induction on k one can show that

(I) the axioms of the intuitionistic propositional calculus belong to $E(J_k)$ for any $k \geqslant 0$.

We shall show this only for the axiom (i_{10}). As it can easily be seen

(1) $\neg A \to (A \to B) \in E(J_0)$.

Let us inductively assume that

(2) $\neg A \to (A \to B) \in E(J_k)$.

Hence by means of lemma 12

(3) $\neg A \to (A \to B) \in E(P_{k+1})$.

Thus

(4) $h(\neg A \to (A \to B)) = b_{P_{k+1}}$

for any mapping $h\colon Var \to P_{k+1}$ satisfying the conditions of definition 5. Let

(5) $\quad h_{k+1}\colon Var \to J_{k+1}\,.$

We shall show that

(6) $\quad$ if $h_{k+1}(A) = b_{k+1} = h_{k+1}(B)$ or if $h_{k+1}(A) \neq b_{k+1} = h_{k+1}(B)$, then $h_{k+1}\big(\neg A \to (A \to B)\big) = b_{k+1}\,.$

Indeed, let us assume that

(6.1) $\quad$ either $h_{k+1}(A) = b_{k+1} = h_{k+1}(B)$
or $h_{k+1}(A) \neq b_{k+1} = h_{k+1}(B)$

and let us observe that by means of definitions 6, 7, 8 and 9

(6.2) $\quad$ if $c = b_m$, then $(a \to_m c) = b_m$

for any $m \geqslant 0$ and for any $a, c \in J_m$. Hence and from (6.1) it follows that

(6.3) $\quad h_{k+1}\big(\neg A \to (A \to B)\big) = b_{k+1}\,.$

We shall further show that

(7) $\quad$ if $h_{k+1}(A) = b_{k+1} \neq h_{k+1}(B)$, then $h_{k+1}\big(\neg A \to (A \to B)\big) = b_{k+1}\,.$

Indeed, let us assume that

(7.1) $\quad h_{k+1}(A) = b_{k+1} \neq h_{k+1}(B)\,.$

Then according to definitions 5, 7 and 9

(7.2) $\quad h_{k+1}(\neg A) = \neg_{k+1} h_{k+1}(A) = \alpha\big(\neg_{P_{k+1}} h_{k+1}(A)\big)\,.$

From (7.1) by means of definition 9 it follows that

(7.3) $\quad \alpha\big(h_{k+1}(B)\big) = h_{k+1}(B)\,.$

From (7.1) and (7.3) by means of definition 7 it follows that

(7.4) $\quad h_{k+1}(A) \to_{k+1} h_{k+1}(B) = \alpha\big(h_{k+1}(A) \to_{P_{k+1}} h_{k+1}(B)\big)$

hence from (7.2) and (4) it follows that

(7.5)
$$\begin{aligned} h_{k+1}\big(\neg A \to (A \to B)\big) &= h_{k+1}(\neg A) \to_{k+1} \big(h_{k+1}(A) \to_{k+1} h_{k+1}(B)\big) = \\ &= \alpha\Big(\neg_{P_{k+1}} h_{k+1}(A) \to_{k+1} \alpha\big(h_{k+1}(A) \to_{P_{k+1}} h_{k+1}(B)\big)\Big) = \\ &= \neg_{P_{k+1}} h_{k+1}(A) \to_{P_{k+1}} \big(h_{k+1}(A) \to_{P_{k+1}} h_{k+1}(B)\big) = \\ &= h\big(\neg A \to (A \to B)\big) = b_{k+1}\,. \end{aligned}$$

At last we shall show that

(8) $\quad$ if $h_{k+1}(A) \neq b_{k+1} \neq h_{k+1}(B)$, then $h_{k+1}\big(\neg A \to (A \to B)\big) = b_{k+1}\,.$

Let us assume that

(8.1) $\quad h_{k+1}(A) \neq b_{k+1} \neq h_{k+1}(B)\,.$

Hence and from definition 6

(8.2) $\quad \alpha\big(h_{k+1}(A)\big) = h_{k+1}(A)$, and $\alpha\big(h_{k+1}(B)\big) = h_{k+1}(B)\,.$

Hence and from definitions 5 and 7

(8.3) $$h_{k+1}\big(\rightharpoondown A\to(A\to B)\big)=h_{k+1}(\rightharpoondown A)\to_{k+1}\big(h_{k+1}(A)\to_{k+1}h_{k+1}(B)\big)=$$
$$=h_{k+1}(\rightharpoondown A)\to_{k+1}\big(h_{k+1}(A)\to_{P_{k+1}}h_{k+1}(B)\big)=$$
$$=h_{k+1}(\rightharpoondown A)\to_{k+1}h(A\to B)\,.$$

From (8.1) by means of definitions 5, 6 and 7

(8.4) $$h_{k+1}(\rightharpoondown A)=\rightharpoondown_{k+1}h_{k+1}(A)=\rightharpoondown_{P_{k+1}}h_{k+1}(A)=h(\rightharpoondown A)$$

Hence

(8.5) $$h_{k+1}(\rightharpoondown A)\to_{k+1}h(A\to B)=h(\rightharpoondown A)\to_{k+1}h(A\to B)\,.$$

Let us observe that according to lemma 13 operation $\to_k$ is normal in J_k. Hence and from lemma 11 operation $\to_{P_{k+1}}$ is normal in P_{k+1} where $P_{k+1}=(J_k)^{k+1}$. Thus according to lemma 8 and definitions 5 and 7

(8.6) $$h(\rightharpoondown A)\to_{k+1}h(A\to B)=h\big(\rightharpoondown A\to(A\to B)\big)=b_{k+1}\,.$$

From (8.3), (8.5), and (8.6) it follows that

(8.7) $$h_{k+1}\big(\rightharpoondown A\to(A\to B)\big)=b_{k+1}\,.$$

From (6), (7) and (8) by means of definition 5 we conclude that

(9) $$\rightharpoondown A\to(A\to B)\in E(J_{k+1})\,.$$

From (1), (2) and (9) it results according to the rule of induction, that for any $k\geqslant 0$

(10) $$\rightharpoondown A\to(A\to B)\in E(J_k)\,.$$

By analogous arguing one can show that each of the remaining axioms (i_1)—(i_9) belongs to $E(J_k)$ for any $k\geqslant 0$.

(II) Now, from lemma 13 and from definition 5 it follows immediately that the set $E(J_k)$ is closed under detachment for any $k\geqslant 0$. This completes the proof.

Let us observe by the way that for any $k\geqslant 0$ the set $E(J_k)$ is trivially closed also under substitution.

4. Jaśkowski's matrix criterion

DEFINITION 10. *For any* $k\geqslant 0$ *we denote by* W_k *the set of all* $A\in S$ *such that*

(i) *either* A *is a theorem of the intuitionistic propositional calculus*

(ii) *or there exist* $B_1, B_2\in S$ *such that* $A=B_1\to B_2$ *and there exists a mapping* $h_k\colon Var\to J_k$ *such that* $h_k(B_1)=b_k\neq h_k(B_2)$.

LEMMA 16. *If* $Q_k\subset W_k$ *and if* $A\in Q_{k+1}$, *then* $[A]_i\in W_k$ *for any* $i\leqslant k+1$.

Proof. Let us assume that

(I) $Q_k \subset W_k$

(II) $A \in Q_{k+1}$

and for any $i \leqslant k+1$

(III) $[A]_i$ is not a theorem of the intuitionistic propositional calculus.

From (II) by means of definition 1 it follows that

(IV) $A = A' \wedge \bigwedge_1^{k+1}(B_i \to b_i) \to a$

where B_i are either simple negations or simple implications. We shall show that

(V) if $B_i = \neg c_i$, then $[A]_i \in W_k$.

Let us assume that

(1) $B_i = \neg c_i$.

Hence and from (III) we conclude according to definition 2 and lemma 4 that

(2) $A' \wedge \bigwedge_1^{k+1}(B_i \to b_i) \to B_i$ is not a theorem of the intuitionistic propositional calculus.

Applying to (2) the well-known Post's theorem about completeness of the classical propositional calculus we obtain according to definition 9 that there exists a mapping $h_0\colon Var \to J_0$ such that

(3) $h_0\big(A' \wedge \bigwedge_1^{k+1}(B_i \to b_i)\big) = b_0 \neq h_0(B_i)$.

We shall define a sequence of mappings $h_m\colon Var \to J_m$ putting

(4) $h_1(a) = h_0(a)$,

$h_m(a) = \{(h_{m-1}(a))_i\}_{i \leqslant m}$

for any $m \leqslant k$ and for any $a \in Var$. Applying lemma 14 to lines (3) and (4) we conclude that

(5) $h_k\big(A' \wedge \bigwedge_1^{k+1}(B_i \to b_i)\big) = b_k \neq h_k(B_i)$.

From (III) and (5) by means of definition 10 it follows that

(6) $[A]_i \in W_k$.

Next we shall show that

(VI) if $B_i = c_i \to d_i$, then $[A]_i \in W_k$.

Let us assume that

(1) $B_i = c_i \to d_i$.

From (1) by means of definition 2 and lemma 1 one can easily conclude that

(2) $[A]_i = A' \wedge \bigwedge_1^{k+1}(B_i \to b_i) \to B_i =$

$= (A' \wedge \bigwedge_1^{k+1-\{i\}}(B_i \to b_i) \wedge \big((c_i \to d_i) \to b_i\big) \to (c_i \to d_i) \leftrightarrow$

$\leftrightarrow (A' \wedge \bigwedge_1^{k+1-\{i\}}(B_i \to b_i) \wedge (d_i \to b_i) \wedge c_i \to d_i)$.

From (II) it also follows that

(3) $(A' \wedge \bigwedge_1^{k+1-\{i\}}(B_i \to b_i) \wedge ((d_i \to b_i) \wedge c_i \to d_i) \in Q_k$.

From (I), (2) and (3) and from definition 10 we conclude that there exists $h_k\colon Var \to J_k$ such that

(4) $h_k(A' \wedge \bigwedge_1^{k+1-\{i\}}(B_i \to b_i) \wedge ((d_i \to b_i) \wedge c_i) = b_k \neq h_k(d_i)$.

Let us observe that the line below

(5) $(A' \wedge \bigwedge_1^{k+1-\{i\}}(B_i \to b_i) \wedge (d_i \to b_i) \wedge c_i) \leftrightarrow A' \wedge \bigwedge_1^{k+1}(B_i \to b_i) \wedge c_i$

holds. Thence and from (4) we obtain by means of lemma 13 and according to definitions 4 and 5 that

(6) $h_k(A' \wedge \bigwedge_1^{k+1}(B_i \to b_i)) = b_k = h_k(c_i)$.

Applying definitions 4 and 5 and lemma 13 to lines (4) and (6), we conclude that

(7) $h_k(c_i \to d_i) = h_k(c_i) \to_k h_k(d_i) \neq b_k$.

From (III), (6) and (7) it follows according to definition 10 that

(8) $[A]_i \in W_k$.

From (V) and (VI) we ultimately obtain that

(VII) $[A]_i \in W_k$.

LEMMA 17. *If* $Q_m^* \subset W_k$ *for any* $m \leqslant k$, *then* $Q_k \subset W_k$.

Proof. Let us assume that for any $m \leqslant k$

(1) $Q_m^* \subset W_k$

(2) $A \in Q_k$

and

(3) A is not a theorem of the intuitionistic propositional calculus.

From (2) and from lemma 6 it follows that for some n

(4) $A \leftrightarrow A_1 \wedge A_2 \wedge \ldots \wedge A_n$

(5) $A = A' \to a$

and for any $j \leqslant n$

(6) $A_j = A_j' \to a$

and there exist $m_j \leqslant k$ such that

(7) $A_j \in Q_{m_j}^*$.

From (3) and (4) it follows that for some $j \leqslant n$

(8) A_j is not a theorem of the intuitionistic propositional calculus while from (1), (6), (7) and (8) it follows according to definition 10 that there exists a mapping

(9) $h_k\colon Var \to J_k$

such that

(10) $h_k(A'_j) = b_k \neq h_k(a)$.

From (9) and (10) we obtain by means of lemma 15 that

(11) $h_k(A'_1 \vee ... \vee A'_j \vee ... \vee A'_n) = b_k$.

From (2) and (5) and from definition 1 and lemma 5 we conclude that

(12) $A'_1 \vee ... \vee A'_j \vee ... \vee A'_n \leftrightarrow A'$.

Hence from (11) and from lemma 15

(13) $h_k(A') = b_k$.

From (3), (10) and (13) it follows according to definition 10 that

(14) $A \in W_k$

and hence and from (2)

(15) $Q_k \subset W_k$,

which completes the proof.

LEMMA 18. *If* $k \leqslant m$, *then* $W_k \subset W_m$.

Proof. Let us assume that

(1) $k \leqslant m$

(2) $A \in W_k$

and

(3) A is not a theorem of the intuitionistic propositional calculus.

From (2) and (3) it follows according to definition 10 that there exist $B_1, B_2 \in S$ such that

(4) $A = B_1 \rightarrow B_2$

and there exists a mapping h_k such that

(5) $h_k: Var \rightarrow J_k$

and

(6) $h_k(B_1) = b_k \neq h_k(B_2)$.

We shall define a mapping $h_m: Var \rightarrow J_m$ thus

(7) if $k = m$, then $h_m(a) = h_k(a)$,

if $k < m$, then (i) $h_{k+1}(a) = \{(h_k(a))_i\}_{i \leqslant 2}$, and

(ii) $h_m(a) = \{(h_{m-1}(a))_i\}_{i \leqslant m}$

for any $a \in Var$. From (6) and (7) it follows according to lemma 14 and definition 8 that

(8) $h_m(B_1) = b_m \neq h_m(B_2)$.

From (3), (4) and (8) it follows according to definition 10, that

(9) $A \in W_m$

which completes the proof.

LEMMA 19. $Q_k \subset W_k$ for any $k \leqslant 0$.

Proof. Let us assume that

(I) $A \in Q_k$

and

(II) A is not a theorem of the intuitionistic propositional calculus.

From (I) and from definition 1 it follows that there exist $n_1, n_2, \ldots, n_6$ such that

(III)
$$A = \bigwedge_1^{n_1} a_i \wedge \bigwedge_1^{n_2} \neg a_i \wedge \bigwedge_1^{n_3}(a_i \to b_i) \wedge \bigwedge_1^{n_4}(a_i \to \neg b_i) \wedge \wedge \bigwedge_1^{n_5}(a_i \to (b_i \to c_i)) \wedge \bigwedge_1^{n_6}(a_i \to b_i \vee c_i) \wedge \bigwedge_1^{k}(B_i \to b_i) \to a = A' \wedge \bigwedge_1^{k}(B_i \to b_i) \to a$$

where B_i is either a simple negation or a simple implication. Let us observe that in view of (I) and (II) we have

(IV) a_i are equal neither a_j nor a

for any $i \leqslant n_1$ and for any $j \leqslant n_2$. We shall prove that

(V) if $k = 0$, then $A \in W_0$.

Indeed, let us assume that

1) $k = 0$.

Hence and from (III) it follows according to lemma 5 that there exist n and $A_1, A_2, \ldots, A_n$ such that for any $i \leqslant n$

(2)
$$A_i = \bigwedge_1^{n_{i,1}} a_i \wedge \bigwedge_1^{n_{i,2}} \neg a_i \wedge \bigwedge_1^{n_{i,3}}(a_i \to b_i) \wedge \bigwedge_1^{n_{i,4}}(a_i \to \neg b_i) \wedge \wedge \bigwedge_1^{n_{i,5}}(a_i \to (b_i \to c_i)) \wedge \bigwedge_1^{n_{i,6}}(a_i \to b_i \vee c_i)$$

(3) $n_{i,1} \geqslant n_1$, $n_{i,2} \geqslant n_2$, $n_{i,3} \geqslant n_3$, $n_{i,4} \leqslant n_4$, $n_{i,5} \leqslant n_5$ and $n_{i,6} \leqslant n_6$

(4) no variable occurring in $\bigwedge_1^{n_{i,1}} a_i$ is equal to any variable occurring in $\bigwedge_1^{n_{i,3}}(a_i \to b_i) \wedge \ldots \wedge \bigwedge_1^{n_{i,6}}(a_i \to b_i \vee c_i)$

and

(5) $A' \leftrightarrow A_1 \vee A_2 \vee \ldots \vee A_n$.

From (5) and from lemma 1 it follows in view of (III) that

(6) $A \leftrightarrow \bigwedge_1^{n}(A_i \to a)$.

From (II) and (6) it follows that there exists $i_0 \leqslant n$ such that

(7) $A_{i_0} \to a$ is not a theorem of the intuitionistic propositional calculus.

From (7) it follows that

(8) no variable occurring in $\bigwedge_1^{n_{i_0,1}} a_i$ is equal to any variable occurring in $\bigwedge_1^{n_{i_0,2}} \neg a_i$.

We shall define a mapping $h_0\colon Var \to J_0$ which takes into account lines (4) and (8) putting

(9) $$\begin{aligned} h_0(a_i) &= 1 \quad \text{for any} \quad i \leqslant n_{i_0,1}\,, \\ h_0(a_i) &= 2 \quad \text{for any} \quad i \leqslant n_{i_0,2}\,, \\ h_0(a_i) &= 2 \quad \text{for any} \quad i \leqslant n_{i_0,3}\,, \\ h_0(a_i) &= 2 \quad \text{for any} \quad i \leqslant n_{i_0,4}\,, \\ h_0(a_i) &= 2 \quad \text{for any} \quad i \leqslant n_{i_0,5}\,, \\ h_0(a_i) &= 2 \quad \text{for any} \quad i \leqslant n_{i_0,6}\,, \\ h_0(a) &= 2\,. \end{aligned}$$

Otherwise, i.e., for any variable b equal to no variable a_i for any $i \leqslant n_{i_0,1}, n_{i_0,2}, \ldots, n_{i_0,6}$ the value $h_0(b)$ is defined arbitrarily. From (9) we obtain by means of definition 9 and lemma 13 that

(10) $$h_0(A_{i_0}) = 1\,.$$

Hence and from the fact that $h_0\colon Var \to J_0$ we obtain by means of lemma 15 and definition 5 that

(11) $$h_0(A_1 \vee \ldots \vee A_{i_0} \vee \ldots \vee A_n) = 1\,.$$

From (5) and (11) it follows according to lemma 15 that

(12) $$h_0(A') = 1$$

while from (II), (9) and (12) we conclude according to definition 10 that

(13) $$A \in W_0\,.$$

Let us inductively assume that

(VI) $Q_m \subset W_m$, for any $m \leqslant k$.

We shall prove that

(VII) if there exists $i \leqslant k$ such that $[A]_i$ is a theorem of the intuitionistic propositional calculus and if $A \in Q_k^*$, then $A \in W_k$.

Let us assume that there exists $i \leqslant k$ such that

(1) $[A]_i$ is a theorem of the intuitionistic propositional calculus and

(2) $A \in Q_k^*$.

From (2) and from definition 2 it follows in view of (III) that

(3) $$[A]_i = A' \wedge \bigwedge_1^k (B_i \to b_i) \to B_i\,.$$

Let us observe that the equivalence

(4) $$A' \wedge \bigwedge_1^k (B_i \to b_i) \leftrightarrow A' \wedge \bigwedge_1^{k-\{i\}} (B_i \to b_i) \wedge b_i$$

holds. Indeed, the implication

$A' \wedge \bigwedge_1^k (B_i \to b_i) \to A' \wedge \bigwedge_1^{k-\{i\}} (B_i \to b_i) \wedge b_i$ is a theorem of the intuitionistic propositional calculus in view of (I). The fact that the converse implication is also a theorem of the intuitionistic propositional calculus is quite evident. Hence from (2) and (3) it follows in view of (III) that

(5) $A \leftrightarrow (A' \wedge \bigwedge_1^{k-\{i\}}(B_i \rightarrow b_i) \wedge b_i \rightarrow a)$

From (II) and (5) it follows that

(6) $A' \wedge \bigwedge_1^{k-\{i\}}(B_i \rightarrow b_i) \wedge b_i \rightarrow a$ is not a theorem of the intuitionistic propositional calculus.

From (2) and from definition 1 we conclude according to (III) that

(7) $(A' \wedge \bigwedge_1^{k-\{i\}}(B_i \rightarrow b_i) \wedge b_i \rightarrow a) \in Q_{k-1}$.

Hence from the inductive hypothesis (VI)

(8) $(A' \wedge \bigwedge_1^{k-\{i\}}(B_i \rightarrow b_i) \wedge b_i \rightarrow a) \in W_{k-1}$.

From (6) and (8) we obtain by means of definition 10 that there exists $h_{k-1}\colon Var \rightarrow J_{k-1}$ such that

(9) $h_{k-1}(A' \wedge \bigwedge_1^{k-\{i\}}(B_i \rightarrow b_i) \wedge b_i) = b_{k-1} \neq h_{k-1}(a)$.

Hence from (4)

(10) $h_{k-1}\big(A' \wedge \bigwedge_1^{k}(B_i \rightarrow b_i)\big) = b_{k-1}$.

For any $a \in Var$ we shall define a mapping $h_k\colon Var \rightarrow J_k$ putting

(11) $h_k(a) = \big\{\big(h_{k-1}(a)\big)_i\big\}_{i \leqslant k}$.

Hence and from lemma 14

(12) $h_k\big(A' \wedge \bigwedge_1^{k}(B_i \rightarrow b_i)\big) = \big\{\big(h_{k-1}(A' \wedge \bigwedge_1^{k}(B_i \rightarrow b_i)\big)_i\big\}_{i \leqslant k}$.

From (9) and (11) and from definition 8

(13) $h_k(a) \neq b_k$.

Finally from (II), (12) and (13) and from definition (10)

(14) $A \in W_k$.

We shall further show that

(VIII) if for any $i \leqslant k$ $[A]_i$ is not a theorem of the intuitionistic propositional calculus and if $A \in Q_k^*$, then $A \in W_k$.

Let us assume that for any $i \leqslant k$

(1) $[A]_i$ is not a theorem of the intuitionistic propositional calculus and

(2) $A \in Q_k^*$.

From (2) and from definition 2 it follows in view of (III) that

(3) $[A]_i = A' \wedge \bigwedge_1^{k}(B_i \rightarrow b_i) \rightarrow B_i$.

From lemma 16 and from inductive hypothesis (VI) as well as from (2) it follows that for any $i \leqslant k$

(4) $[A]_i \in W_{k-1}$.

Hence and from (1) it follows according to definition 10 that there exist mappings $h_{k-1,i}\colon Var \to J_{k-1}$ such that

(5) $$h_{k-1,i}\big(A' \wedge \bigwedge_1^k (B_i \to b_i)\big) = b_{k-1} \neq h_{k-1,i}(B_i)\,.$$

Taking into account the mappings $h_{k-1,i}$ we shall define a mapping $g\colon Var \to P_k$ putting

(6) $$g(a) = \{h_{k-1,i}(a)\}_{i \leqslant k}$$

for any $a \in Var$ where of course $P_k = (J_{k-1})^k$. Further, taking into account the mapping g we shall define a mapping $h_k\colon Var \to J_k$ putting

(7) $$h_k(a) = \begin{cases} g(a_i), & \text{if there exists } i \leqslant n_1 \text{ such that } a = a_i \\ a\big(g(a)\big), & \text{otherwise} \end{cases}$$

where a is an arbitrary variable and where n_1 is the number defined in line (III). Applying lemma 12 to lines (5) and (6) we conclude according to definitions 5 and 8 that

(8) $$g\big(A' \wedge \bigwedge_1^k (B_i \to b_i)\big) = b_{P_k}\,.$$

Now we shall show that

(*) $$h_k\big(A' \wedge \bigwedge_1^k (B_i \to b_i)\big) = g\big(A' \wedge \bigwedge_1^k (B_i \to b_i)\big)$$

considering successively the syntactic structure of all the factors occurring in the conjunction $A' \wedge \bigwedge_1^k (B_i \to b_i)$. From definition (7) it follows immediately that for any $i \leqslant n_1$

(9) $$h_k(a_i) = g(a_i)\,.$$

From (IV) and (2) we obtain that for any $i \leqslant n_1$ and for any $j \leqslant n_2, n_3, \ldots, n_6, k$

(10) a_i is not equal to a_j.

From (7) and (10) it follows that for any $i \leqslant n_2, n_3, \ldots, n_6, k$

(11) $$h_k(a_i) = a\big(g(a_i)\big)\,.$$

From (11) and from definitions 5 and 7 it follows that for any $i \leqslant n_2$

(12) $$h_k(\neg a_i) = \neg_k h_k(a_i) = \neg_k a\big(g(a_i)\big) = \neg_{P_k} g(a_i) = g(\neg a_i)\,.$$

From (11) and from definitions 5 and 7 it follows that for any $i \leqslant n_3$

(13) $$h_k(a_i \to b_i) = h_k(a_i) \to_k h_k(b_i) = a\big(g(a_i)\big) \to_k a\big(g(b_i)\big) = \\ = g(a_i) \to_{P_k} g(b_i) = g(a_i \to b_i)\,.$$

From (11) and from definitions 5 and 7 it follows that for any $i \leqslant n_4$

(14.1) $$h_k(a_i \to \neg b_i) = h_k(a_i) \to_k \neg_k h_k(b_i) = \\ = a\big(g(a_i)\big) \to_k \neg_k a\big(g(b_i)\big) = a\big(g(a_i)\big) \to_k \neg_{P_k} g(b_i)\,.$$

According to definitions 5 and 7

(14.2) if $g(\neg b_i) = b_{P_k}$, then $a\big(g(a_i)\big) \to_k \neg_{P_k} g(b_i) =$ $= g(a_i) \to_{P_k} g(\neg b_i) = g(a_i \to \neg b_i)\,.$

According to definitions 5 and 7 we obtain also that

(14.3) if $g(\rightharpoondown b_i) \neq b_{P_k}$, then $a(g(a_i)) \rightarrow_k \rightharpoondown_{P_k} g(b_i) =$
$= g(a_i) \rightarrow_{P_k} g(\rightharpoondown b_i) = g(a_i \rightarrow \rightharpoondown b_i)$.

From (14.1), (14.2) and (14.3) it follows that for any $i \leqslant n_4$

(14) $h_k(a_i \rightarrow \rightharpoondown b_i) = g(a_i \rightarrow \rightharpoondown b_i)$.

From (11) and from definitions 5 and 7 it follows that for any $i \leqslant n_5$

(15.1) $h_k(a_i \rightarrow (b_i \rightarrow c_i)) = h_k(a_i \rightarrow_k (h_k(b_i) \rightarrow_k h_k(c_i)) =$
$= a(g(a_i)) \rightarrow_k (a(g(b_i)) \rightarrow_k a(g(c_i))) =$
$= a(g(a_i)) \rightarrow_k (g(b_i) \rightarrow_{P_k} (g(c_i)) = a(g(a_i)) \rightarrow_k g(b_i \rightarrow c_i)$.

According to definitions 5 and 7

(15.2) if $g(b_i \rightarrow c_i) = b_{P_k}$, then $a(g(a_i)) \rightarrow_k g(b_i \rightarrow c_i) =$
$= g(a_i) \rightarrow_{P_k} g(b_i \rightarrow c_i) = g(a_i \rightarrow (b_i \rightarrow c_i))$.

According to definitions 5 and 7

(15.3) if $g(b_i \rightarrow c_i) \neq b_{P_k}$, then $a(g(a_i) \rightarrow_k g(b_i \rightarrow c_i) =$
$= g(a_i) \rightarrow_{P_k} g(b_i \rightarrow c_i) = g(a_i \rightarrow (b_i \rightarrow c_i))$

From (15.1), (15.2) and (15.3) it follows that for any $i \leqslant n_5$

(15) $h_k(a_i \rightarrow (b_i \rightarrow c_i)) = g(a_i \rightarrow (b_i \rightarrow c_i))$.

From (11) and from definitions 5 and 7 it follows that for any $i \leqslant n_6$.

(16) $h_k(a_i \rightarrow b_i \vee c_i) = h_k(a_i) \rightarrow_k (h_k(b_i) \vee_k h_k(c_i)) =$
$= a(g(a_i)) \rightarrow_k (a(g(b_i) \vee_k a(g(c_i))) =$
$= a(g(a_i)) \rightarrow_k a(g(b_i) \vee_{P_k} g(c_i)) =$
$= a(g(a_i)) \rightarrow_k a(g(b_i \vee c_i)) =$
$= g(a_i) \rightarrow_{P_k} g(b_i \vee c_i) = g(a_i \rightarrow b_i \vee c_i)$.

Let us assume that for $i \leqslant k$

(17.1) $B_i = \rightharpoondown a_i$.

From (5), (6) and (17.1) it follows that

(17.2) $g(\rightharpoondown a_i) \neq b_{P_k}$.

Hence and from definition 6

(17.3) $g(\rightharpoondown a_i) = a(g(\rightharpoondown a_i))$.

From (11) and (17.3) it follows by means of definitions 5, 6 and 7 that

(17.4) $h_k(B_i \rightarrow b_i) = h_k(\rightharpoondown a_i \rightarrow b_i) = \rightharpoondown_k h_k(a_i) \rightarrow_k h_k(b_i) =$
$= \rightharpoondown_k a(g(a_i)) \rightarrow_k a(g(b_i)) = \rightharpoondown_{P_k} g(a_i) \rightarrow_k a(g(b_i)) =$
$= g(\rightharpoondown a_i) \rightarrow_k a(g(b_i)) = a(g(\rightharpoondown a_i) \rightarrow_k a(g(b_i)) =$
$= g(\rightharpoondown a_i) \rightarrow_{P_k} (g(b_i)) = g(\rightharpoondown a_i \rightarrow b_i) = g(B_i \rightarrow b_i)$.

Thus, for any $i \leqslant k$

(17) if $B_i = \rightharpoondown a_i$, then $h_k(B_i \rightarrow b_i) = g(B_i \rightarrow b_i)$.

Let us further assume that for $i \leqslant k$

(18.1) $B_i = a_i \to c_i$.

From (5), (6) and (18.1) it follows that

(18.2) $g(a_i \to c_i) \neq b_{P_k}$.

Hence and from definition 6

(18.3) $g(a_i \to c_i) = a\big(g(a_i \to c_i)\big)$.

From (11) and (18.3) it follows by means of definitions 5, 6 and 7 that

(18.4)
$$\begin{aligned} h_k(B_i \to b_i) &= h_k\big((a_i \to c_i) \to b_i\big) = \\ &= \big(h_k(a_i) \to_k h_k(c_i)\big) \to_k h_k(b_i) = \\ &= \Big(a\big(g(a_i)\big) \to_k a\big(g(c_i)\big)\Big) \to_k a\big(g(b_i)\big) = \\ &= \big(g(a_i) \to_{P_k} g(c_i)\big) \to_k a\big(g(b_i)\big) = \\ &= g(a_i \to c_i) \to_k a\big(g(b_i)\big) = a\big(g(a_i \to c_i)\big) \to_k a\big(g(b_i)\big) = \\ &= g(a_i \to c_i) \to_{P_k} g(b_i) = g\big((a_i \to c_i) \to b_i\big) = g(B_i \to b_i). \end{aligned}$$

Thus, for any $i \leqslant k$

(18) if $B_i = a_i \to c_i$, then $h_k(B_i \to b_i) = g(B_i \to b_i)$.

From (17) and (18) it follows that for any $i \leqslant k$

(19) $h_k(B_i \to b_i) = g(B_i \to b_i)$.

From (9), (12)–(16) and from (19) according to definition 5 the equality (*) follows, i.e.,

(20) $h_k\big(A' \wedge \bigwedge_1^k (B_i \to b_i)\big) = g\big(A' \wedge \bigwedge_1^k (B_i \to b_i)\big)$.

Hence and from (8) it follows in view of definition 7 that

(21) $h_k\big(A' \wedge \bigwedge_1^k (B_i \to b_i)\big) = b_k$.

Let us further observe that according to (IV), (7) and definition 6

(22) $h_k(a) = a\big(g(a)\big) \neq b_k$.

From (II), (21) and (22) and from definition 10 it follows ultimately that

(23) $A \in W_k$.

From (VII) and (VIII) it follows that

(IX) $Q_k^* \subset W_k$.

From (VI), from lemma 18 and from definition of the class Q_m^* it follows that

(X) $Q_m^* \subset W_k$, for any $m < k$.

Hence and from (IX)

(XI) $Q_m^* \subset W_k$, for any $m \leqslant k$.

Applying to (XI) lemma 17 we obtain that

(XII) $Q_k \subset W_k$.

From (V), (VI) and from (XII) it follows according to the rule of induction that

(XIII) $Q_k \subset W_k$, for any $k \geqslant 0$

which completes the proof.

Let us notice quite obvious

LEMMA 20. *For an arbitrary matrix M if $E(M)$ contains the set of all theorems of the intuitionistic propositional calculus and if $A \in E(M)$, then the condition stating that $B \Leftrightarrow A$ entails the condition stating that $B \in E(M)$.*

LEMMA 21. *If $A \in E(J_k)$ for any $k \geqslant 0$, then A is a theorem of the intuitionistic propositional calculus.*

Proof. Let us assume that for any $k \geqslant 0$

(1) $A \in E(J_k)$

and let us indirectly suppose that

(2) A is not a theorem of the intuitionistic propositional calculus.

According to lemma 7 there exist k_0 and

(3) $B \in Q^*_{k_0}$

such that

(4) $A \Leftrightarrow B$.

From (3) it follows that

(5) $B \in Q_{k_0}$

and hence and from lemma 19

(6) $B \in W_{k_0}$.

From (2) and (4) it follows that

(7) B is not a theorem of the intuitionistic propositional calculus

hence and from (6) it results in view of definition 10 that

(8) $B \notin E(J_{k_0})$.

According to lemma 15

(9) $E(J_{k_0})$ contains the set of all theorems of the intuitionistic propositional calculus.

Hence, from (4) and (7) it follows according to lemma 20 that

(10) $A \notin E(J_{k_0})$

contrary to line (1).

From lemmas 15 and 21 there results immediately:

JAŚKOWSKI'S MATRIX CRITERION. *For any* $A \in S$ *the condition stating that* A *is a theorem of the intuitionistic propositional calculus is equivalent to the condition stating that* $A \in E(J_k)$, *for any* $k \geqslant 0$.

Thus, the sequence $\{J_k\}_{k\geqslant 0}$ of Jaśkowski's matrices, defined in definition 9, is a characteristic sequence for the intuitionistic propositional calculus. Now let us consider another sequence $\{J'_k\}_{k\geqslant 0}$ of matrices, defined by means of the equalities:

(i) $J'_0 = J_0 = L_1$

(ii) $J'_{k+1} = \Gamma((J'_k)^2)$.

C. G. Mckay has stated that the sequence $\{J'_k\}_{k\geqslant 0}$ is also characteristic for intuitionistic propositional calculus (cf. [3]). But it is not true (cf. [4]).

References

[1] HEYTING A.: *Die formalen Regeln der intuitionistischen Logik.* Sitzungsberichte der Preussischen Akademie der Wissenschaften. Physikalisch-mathematische Klasse, 1930, 42–56.

[2] JAŚKOWSKI S.: *Recherches sur le système de la logique intuitionniste.* Actes du Congrès International de Philosophie Scientifique VI. Philosophie des Mathématiques. Actualités Scientifiques et Industrielles, 393(1936), 58–61.

[3] MCKAY C. G.: *A note on the Jaśkowski sequence.* Zeitschrift für mathematische Logik und Grundlagen der Mathematik, 13(1967), 95–96.

[4] ONO, HIROAKIRA: *Kripke models and intermediate logics.* Publications of the Research Institute for Mathematical Sciences, 6(1971), 461–476, Kyoto.

[5] ROSE G. F.: *Propositional calculus and realizability.* Transactions of the American Mathematical Society, 75(1953), 1–19.

[6] SURMA S. J.: *Twierdzenia o dedukcji niewprost.* Studia Logica, 20(1967), 151–160.

[7] SURMA S. J.: *A historical survey of the significant methods for proving Post's theorem about the completeness of the classical propositional calculus.* This volume.

ANDRZEJ WROŃSKI

AXIOMATIZATION OF THE IMPLICATIONAL GÖDEL'S MATRICES BY KALMAR'S METHOD

According to A. Church (cf. [2], p. 164) the adaptation of the well-known Kalmar's method of proving completeness (cf. [7]) to the case of the implicational fragment of the classical propositional calculus is due to L. Henkin [1]. In this paper the result of Henkin is generalized to the cases of certain implicational intermediate calculi. It is shown that provability in these calculi is equivalent to validity in corresponding implicational Gödel's matrices [2]. In the paper the parenthesis-free notation of Łukasiewicz [3] and the well-known set-theoretical notation [4] is used. The set $\{1, 2, \ldots\}$ of natural numbers is denoted by N. The elements of the set $Z = \{z_1, z_2, \ldots\}$ are called variables. The elements of the smallest set F such that $Z \subset F$ and fab $\in F$ for every $a, b \in F$ are called formulas. The expression $\underline{ai/b}$ denotes the formula resulting by substitution of the formula b for the variable z_i in the formula a. The symbol $\underline{a}$ denotes the smallest set containing the formula a and closed under substitution. The symbol LH denotes the set of theorems of the implicational propositional calculus of Hilbert (cf. [6]) whereas the symbol $\vdash$ denotes the relation of derivability by means of LH and detachment. For every $n \in N$ the symbol G^n denotes the formula such that $G^1 = z_1$, $G^{n+1} = fffz_n z_{n+1} z_1 G^n$ (cf. [12]). The symbol $\underset{n}{\vdash}$ denotes the relation of derivability by means of $LH \cup \underline{G^n}$ and detachment. The symbol LG^n

[1] For numbers in the square brackets cf. p. 131.

[2] It is familiar in logical literature to apply the name „n-th Gödel's matrix" to the n-th element of the sequence of matrices defined by K. Gödel in [3]. The second Gödel's matrix is sometimes called Heyting's matrix, because A. Heyting defined it in [4]. This matrix was axiomatized already in 1938 by J. Łukasiewicz (cf. [8], 91). The axiomatization of the whole sequence of Gödel's matrices is due to I. Thomas (cf. [12]).

[3] Cf. [9].

[4] Cf. for instance [10].

denotes the set $\{a \in F : \vdash_n a\}$. In order to avoid unnecessary complications we shall not undertake the exact formalization of the proofs. Above all the deduction theorem (cf. [4], [11]) and many theorems of LH will be freely used. We shall establish the following series of lemmas to be used later.

(0) LG^n is closed under substitution.

Proof. If $a \in LG^n$, then $fg_1fg_2...fg_ka \in LH$ for certain $g_1, g_2, ..., g_k \in \underline{G}^n$. It is well-known that LH is closed under substitution. Hence we have $\underline{fg_1fg_2...fg_ka} \subset LH \subset LG^n$ and $\underline{g_1}, \underline{g_2}, ..., \underline{g_k} \subset \underline{G}^n \subset LG^n$. Thus, $\underline{a} \subset LG^n$ since LG^n is closed under detachment. Q. E. D.

(1) $\vdash_n fffbcafffcbaa$[5]

Proof.

1	$ffbca$	assumption
2	$ffcba$	assumption
3	fba	2.
4	$ffaca$	1, 3.
5	a	1, 2, 4, $\underline{G}^n$. Q. E. D.

(2) $\vdash_n ffffcbbafffcbaa$

Proof.

1	$fffcbba$	assumption
2	$ffcba$	assumption
3	$fffcbcffcbb$	
4	$fffcbca$	1, 3.
5	$ffcfcba$	2.
6	a	4, 5, (1). Q. E. D.

Let P^k be the abbreviation for $ffz_kz_{k+1}z_1$, so P^k stands for k-th antecedent of any formula G^n such that $k < n$.

(3) If $1 \leqslant i < j \leqslant m$, then $\vdash_n ffz_iz_jG^m$

Proof by induction on $j-i$.

I	$j-i = 1$	assumption
1	$1 \leqslant i < j \leqslant m$	assumption
2	fz_iz_j	assumption

---

[5] This formula is an axiom for the implicational fragment of the Dummett's calculus LC (cf. [1]).

3	P^k for every $1 \leqslant k < m$	assumption
4	$1 \leqslant i < m$	1.
5	P^i	3, 4.
6	$j = i+1$	I.
7	$fz_i z_{i+1}$	2, 6.
8	z_1	5, 7.
II	$j-i > 1$ and (3) holds for every j', i' such that $j'-i' < j-i$	assumption
1	$1 \leqslant i < j \leqslant m$	assumption
2	$fz_i z_j$	assumption
3	$i+1 < j$	II.
4	$1 \leqslant i+1 < j \leqslant m$	1, 3.
5	$1 \leqslant i < i+1 \leqslant m$	1, 3.
6	$ffz_{i+1} z_j G^m$	4, II.
7	$ffz_i z_{i+1} G^m$	5, II.
8	$ffz_{i+1} z_i G^m$	2, 6.
9	G_m	7, 8, (1). Q. E. D.

(4) $\vdash fffbG^nG^nfffabG^nG^n$

Proof. Trivial.

(5) $\vdash fffaG^nG^nff\underline{G^n i/b}G^nf\underline{G^n i/fa}bG^n$

Proof.

1	$ffaG^nG^n$	assumption
2	$f\underline{G^n i/b}G^n$	assumption
3	$\underline{G^n i/fa}b$	assumption
4	$ffbfabfffabbf\underline{G^n i/fa}b\underline{G^n i/b}$	
5	$fffabbG^n$	2, 3, 4.
6	faG^n	5.
7	G^n	1, 6. Q. E. D.

(6) $\vdash_{n+1} ff\underline{G^n n/a}G^nfffabG^nG^n$

Proof by considering the following cases: I, II.

I	$n = 1$	assumption
1	$f\underline{G^n n/a}G^n$	assumption
2	$ffabG^n$	assumption
3	faz_1	1, I.
4	$ffabz_1$	2, I.

5	ffz_1bz_1	3, 4.
6	z_1	5, $\underline{G^2}$.
II	$n>1$	assumption
1	$f\underline{G^n n/a}G^n$	assumption
2	$ffabG^n$	assumption
3	P^k for every $1\leqslant k<n$	assumption
4	$ffabz_1$	2, 3.
5	$fffabz_1\underline{G^n n/a}$	$\underline{G^{n+1}}$, II.
6	G^n	1, 4, 5.
7	z_1	3, 6. Q. E. D.

(7) If $1\leqslant i\leqslant n$, then

$\vdash_{n+1} ff\underline{G^n i/a}G^nff\underline{G^n i/b}G^nfffabG^nG^n$

Proof by considering the following cases: I, II, III.

I	$1=i<n$	assumption
1	$f\underline{G^n i/a}G^n$	assumption
2	$f\underline{G^n i/b}G^n$	assumption
3	$ffabG^n$	assumption
4	P^k for every $1\leqslant k<n$	assumption
5	$f\underline{G^n 1/a}z_1$	1, 4, I.
6	$f\underline{G^n 1/b}z_1$	2, 4, I.
7	$fa\underline{G^n 1/a}$	
8	$2\leqslant n$	I.
9	$ffbz_2\underline{G^n 1/b}$	8, (3).
10	faz_1	5, 7.
11	$ffabz_1$	3, 4.
12	ffz_1bz_1	10, 11.
13	$ffbz_2z_1$	6, 9.
14	z_1	4, 12, 13, $\underline{G^{n+1}}$
II	$1<i<n$	assumption
1	$f\underline{G^n i/a}G^n$	assumption
2	$f\underline{G^n i/b}G^n$	assumption
3	$ffabG^n$	assumption
4	P^k for every $1\leqslant k<n$	assumption
5	$1\leqslant i-1<i\leqslant n$	II.
6	$1\leqslant i<i+1\leqslant n$	II.

7	$ffz_{i-1}a\underline{G^n i/a}$	5, (3).
8	$ffbz_{i+1}\underline{G^n i/b}$	6, (3).
9	$ffz_{i-1}az_1$	1, 4, 7.
10	$ffabz_1$	3, 4.
11	$ffbz_{i+1}z_1$	2, 4, 8.
12	z_1	4, 9, 10, 11, $\underline{G^{n+1}}$.
III	$1 \leqslant i = n$	assumption
1	$f\underline{G^n i/a}G^n$	assumption
2	$f\underline{G^n i/b}G^n$	assumption
3	$f\underline{G^n n/a}G^n$	1, III.
4	$fffabG^nG^n$	3, (6). Q. E. D.

(8) If $1 \leqslant i < j \leqslant n$, then
$\vdash_{n+1} ff\underline{G^n i/a}G^n ff\underline{G^n j/b}G^n f\underline{G^n j/fab}G^n$

Proof.

1	$1 \leqslant i < j \leqslant n$	assumption
2	$f\underline{G^n i/a}G^n$	assumption
3	$f\underline{G^n j/b}G^n$	assumption
4	$\underline{G^n j/fab}$	assumption
5	P^k for every $1 \leqslant k < n$	assumption
6	$1 \leqslant i < i+1 \leqslant n$	1.
7	$ffaz_{i+1}\underline{G^n i/a}$	6, (3).
8	$ffz_i b\underline{G^n j/b}$	1, (3).
9	$ffz_i bz_1$	3, 5, 8.
10	$ffabffz_i az_1$	9.
11	$ffaz_{i+1}z_1$	2, 5, 7.
12	$ffabz_1$	5, 10, 11, $\underline{G^{n+1}}$.
13	$ffbfabfffabbf\underline{G^n j/fab}\underline{G^n j/b}$	
14	$fffabb\underline{G^n j/b}$	4, 13.
15	$ffffabbz_1$	3, 5, 14.
16	z_1	12, 15, (2). Q. E. D.

(9) If $1 \leqslant i < j \leqslant n$, then
$\vdash_{n+1} ff\underline{G^n i/a}G^n ff\underline{G^n j/b}G^n fffbaG^nG^n$

Proof.

1	$1 \leqslant i < j \leqslant n$	assumption
2	$f\underline{G^n i/a}G^n$	assumption

3	$f\underline{G^n j/b}G^n$	assumption
4	$ffbaG^n$	assumption
5	$f\underline{G^n j/fab}G^n$	1, 2, 3, (8).
6	$ffz_i fab\underline{G^n j/fab}$	1, (3).
7	$ffz_i fabG^n$	5, 6.
8	$ffabG^n$	7.
9	G^n	4, 8, (1). Q. E. D.

(10) $\vdash_{n+1} fffaG^nG^nffG^naa$

Proof.

1	$ffaG^nG^n$	assumption
2	fG^na	assumption
3	$ffz_iz_{i+1}G^n$ for every $1 \leqslant i < n$	(3).
4	$ffz_iz_{i+1}a$ for every $1 \leqslant i < n$	2, 3.
5	$ffaG^na$	1, 2.
6	fz_1G^n	
7	$ffaz_1a$	5, 6.
8	a	4, 7, $\underline{G^{n+1}}$. Q. E. D.

(11) $\vdash f\underline{G^n1/a}\ldots f\underline{G^nn/a}ffaG^nG^n$

Proof.

1	$\underline{G^n1/a}$	assumption
2	$\underline{G^ni/a}$ for every $1 < i < n$	assumption
3	$\underline{G^nn/a}$	assumption
4	faG^n	assumption
5	P^k for every $1 \leqslant k < n$	assumption
6	$f\underline{P^{i-1}i/a}f\underline{P^ii/a}z_1$ for every $1 < i < n$ because for every l such that $l \neq i$ and $l \neq i-1$ there is $\underline{P^li/a} = P^l$	2, 5,
7	$fP^iffz_ia\underline{P^ii/a}$	
8	$f\underline{P^{i-1}i/a}ffz_iaz_1$ for every $1 < i < n$	5, 6, 7.
9	$f\underline{P^{i-1}i/a}\underline{P^ii+1/a}$ for every $1 < i < n$	8.
10	$ffaz_1ffz_1af\underline{G^n1/a}G^n$	
11	faz_1	4, 5.
12	ffz_1az_1	1, 5, 10, 11.
13	$\underline{P^12/a}$	12.
14	$\underline{P^{n-1}n/a}$ if $n > 1$	9, 13.

15	z_1 if $n > 1$	3, 5, 14.
16	z_1 if $n = 1$	1, 11.
17	z_1	15, 16. Q. E. D.

(12) $\{ff\underline{G^n1/a}G^nb, \ldots, ff\underline{G^nn/a}G^nb\} \underset{n+1}{\vdash} ffffaG^nG^nbb$

Proof.

1	$ff\underline{G^ni/a}G^nb$ for every $i = 1, \ldots, n$	assumption
2	$fffaG^nG^nb$	assumption
3	$ffbG^nff\underline{G^ni/a}G^nG^n$ for every $i = 1, \ldots, n$	1.
4	$ffbG^nff\underline{G^n1/a}\ldots f\underline{G^nn/a}G^nG^n$	3.
5	$f\underline{G^n1/a}\ldots f\underline{G^nn/a}ffaG^nG^n$	(11).
6	$ffbG^nffffaG^nG^nG^nG^n$	4, 5.
7	$ffbG^nfffaG^nG^nG^n$	2.
8	$ffbG^nG^n$	6, 7.
9	ffG^nbb	8, 10.
10	fG^nb	2.
11	b	9, 10. Q. E. D.

The content of the matrix $\mathfrak{M} = \langle U^{\mathfrak{M}}, W^{\mathfrak{M}}, f^{\mathfrak{M}} \rangle$ is denoted by $E(\mathfrak{M})$ and the value of formula $a \in F$ with respect to the valuation $v: Z \to U^{\mathfrak{M}}$ is denoted by a^v. The matrix $\mathfrak{G}^n = \langle U^n, W^n, f^n \rangle$ is called n-th Gödel's matrix [6], if:

1 $U^n = \{1, \ldots, n\}$

2 $f^n: U^n \times U^n \to U^n$, $W^n = \{1\}$
$f^n(i, j) = 1$ if $i \geqslant j$
$f^n(i, j) = j$ if $i < j$

The next lemma will be stated without proof.

(13) $E(\mathfrak{G}^{n+1}) \subset E(\mathfrak{G}^n)$
$LG^n \subset E(\mathfrak{G}^n)$.

For every valuation $v: Z \to U^{n+1}$ the symbol v^{n+1} denotes the function: $F \to F$ such that:

1	$v^{n+1}(a) = ffaG^nG^n$	if $a^v = 1$
2	$v^{n+1}(a) = f\underline{G^na^v - 1/a}G^n$	if $a^v > 1$.

On the basis of the conventions above one can obtain:

(14) If $v: Z \to U^{n+1}$ and $b_1, \ldots, b_k$ are all the variables occurring in formula a, then
$\{v^{n+1}(b_1), \ldots, v^{n+1}(b_k)\} \underset{n+1}{\vdash} v^{n+1}(a)$.

---

[6] In the Gödel's paper [3] the degenerate matrix $\mathfrak{G}^1$ is omitted.

Proof by induction on the number of occurrences of f in a may be sketched as follows. If f does not occur in a, the lemma is trivial. From lemmas (4), ..., (9) it follows that $\vdash_{n+1} fv^{n+1}(a)fv^{n+1}(b)v^{n+1}(fab)$. Thus, if the lemma holds for a and for b, then it also holds for fab. Q. E. D.

(15) If $b_1, \ldots, b_k$ are variables and $\{v^{n+1}(b_1), \ldots, v^{n+1}(b_k)\} \vdash_{n+1} a$ for every $v: Z \to U^{n+1}$, then $\vdash_{n+1} a$.

Proof by induction on k.

I	$k = 1$	assumption
1	$b_1, \ldots, b_k$ are variables	assumption
2	$\{v^{n+1}(b_1), \ldots, v^{n+1}(b_k)\} \vdash_{n+1} a$ for every $v: Z \to U^{n+1}$	assumption
3	$\vdash_{n+1} fv^{n+1}(b_1)a$ for every $v: Z \to U^{n+1}$	2, I.
4	$\vdash_{n+1} fv^{n+1}(b_1)a$ and $b_1^v = 1$	1, 3.
5	$\vdash_{n+1} fv^{n+1}(b_1)a$ and $b_1^v = 2, \ldots, n+1$	1, 3.
6	$\vdash_{n+1} fffb_1G^nG^na$	4.
7	$\vdash_{n+1} ff\underline{G^ni/b_1}G^na$ for every $i = 1, \ldots, n$	5.
8	$\vdash_{n+1} ffffb_1G^nG^naa$	7, (12).
9	$\vdash_{n+1} a$	6, 8.
II	$k > 1$ and (15) holds for every $k' < k$	assumption
1	$b_1, \ldots, b_k$ are variables	assumption
2	$\{v^{n+1}(b_1), \ldots, v^{n+1}(b_k)\} \vdash_{n+1} a$ for every $v: Z \to U^{n+1}$	assumption
3	$\{v^{n+1}(b_1), \ldots, v^{n+1}(b_{k-1}), x^{n+1}(b_k)\} \vdash_{n+1} a$ for every $v, x: Z \to U^{n+1}$	1, 2.
4	$\{v^{n+1}(b_1), \ldots, v^{n+1}(b_{k-1})\} \vdash_{n+1} fx^{n+1}(b_k)a$ for every $v, x: Z \to U^{n+1}$	3.
5	$\vdash_{n+1} fx^{n+1}(b_k)a$ for every $x: Z \to U^{n+1}$	1, 4, II.
6	$\vdash_{n+1} a$.	1, 5, II. Q. E. D.

An easy consequence of (14) and (15) is:

(16) If $a \in E(\mathfrak{G}^{n+1})$, then $\vdash_{n+1} ffaG^nG^n$.

This fact enables us to prove the following completeness theorem for LG^n.

THEOREM $LG^n = E(\mathfrak{G}^n)$.

Proof. The inclusion $LG^n \subset E(\mathfrak{G}^n)$ follows from (13). Since for every $n \in N$ the set of variables z_i such that $i > n$ is infinite and LG^n is shown (lemma (0)) to be closed under substitution, then to prove the converse inclusion it suffices to establish that

(*) If for every $i \leqslant n$ the variable z_i does not occur in the formula a and $a \in E(\mathfrak{G}^n)$, then $\underset{n+1}{\vdash} a$.

Proving (*) by induction on n we consider only the induction step because the case $n = 1$ is trivial.

1	(*) holds for n	assumption
2	for every $i \leqslant n+1$ variable z_i does not occur in the formula a	assumption
3	$a \in E(\mathfrak{G}^{n+1})$	assumption
4	for every $i \leqslant n$ variable z_i does not occur in the formula a	2.
5	$a \in E(\mathfrak{G}^n)$	3, (13).
6	$\underset{n}{\vdash} a$	1, 4, 5.
7	$\vdash fg_1 \ldots fg_k a$ for certain $g_1, \ldots, g_k \in \underline{G^n}$	6.
8	$\underset{+1}{\vdash} ffaG^nG^n$	3, (16).
9	$\underset{n+1}{\vdash} ffG^naa$	8, (10).
10	$\underset{n+1}{\vdash} ffg_iaa$ for every $i = 1, \ldots, k$	2, 7, 9, (0).
11	$\underset{n+1}{\vdash} ffg_1 \ldots fg_k aa$	10.
12	$\underset{n+1}{\vdash} a$	7, 11. Q. E. D.

References

[1] BULL R. A.: *The implicational fragment of Dummett's LC.* Journal of Symbolic Logic, 27(1962), 189–194.

[2] CHURCH A.: *Introduction to mathematical logic.* Vol. I. Princeton 1956.

[3] GÖDEL K.: *Zum intuitionistischen Aussagenkalkül.* Akademie der Wissenschaften in Wien, Mathematisch-naturwissenschaftliche Klasse, Anzeiger, 69(1932), 65–66.

[4] HERBRAND J.: *Recherches sur la théorie de la démonstration.* Travaux de la Société des Sciences et des Lettres de Varsovie, Classe III, 33(1930).

[5] Heyting A.: *Die formalen Reglen der intuitionistischen Logik.* Sitzungsberichte der Preussischen Akademie der Wissenschaften, Physikalisch--mathematische Klasse, 1930, 42–56.
[6] Hilbert D., Bernays P.: *Grundlagen der Mathematik.* Vol. I. Berlin 1934.
[7] Kalmar L.: *Über die Axiomatisierbarkeit des Aussagenkalküls.* Acta Scientiarum Mathematicarum, 7(1935), 222–243.
[8] Łukasiewicz J.: *Die Logik und das Grundlagenproblem.* In: Les entretiens de Zürich sur les fondements et la méthode des sciences mathématiques 6–9 XII 1938. Zürich 1941, 82–100.
[9] Łukasiewicz J.: *Elementy logiki matematycznej.* Warszawa 1958.
[10] Rasiowa H.: *Wstęp do matematyki współczesnej.* Warszawa 1968.
[11] Tarski A.: *Über einige fundamentale Begriffe der Metamathematik.* Comptes Rendus des Séances de la Société des Sciences et des Lettres de Varsovie, Classe III, 23(1930), 22–29.
[12] Thomas I.: *Finite Limitations of Dummett's LC.* Notre Dame Journal of Formal Logic, 3(1962), 170–174.

ANDRZEJ WROŃSKI

A CONTRIBUTION TO THE HISTORY OF INVESTIGATIONS INTO THE INTERMEDIATE PROPOSITIONAL CALCULI [1]

The aim of this paper is a short review of some results concerned with the problem of axiomatizability of the lattices with relative pseudo-complement and zero [2] which in the sequel will be called simply lattices. The investigations into the propositional calculi intermediate between the intuitionistic and the classical calculus can be reduced to the investigations into such lattices in view of the connections discovered by Stone [16] and Tarski [17]. In this paper the well-known symbolism of mathematical logic and the following notations will be used. The symbols I, K, A, N stand for implication, conjunction, disjunction and negation, respectively [3]. The symbol Ax denotes a system of axioms of the intuitionistic propositional calculus and the symbol Cn denotes the consequence operation on the basis Ax by means of substitution and detachment. The symbol $E(\mathfrak{M})$ denotes the content of the lattice $\mathfrak{M}$ and the symbol $\mathfrak{B}$ will be reserved for the classical lattice, i. e., the two-element Boolean algebra. The Cartesian product of the lattices $\mathfrak{M}$ and $\mathfrak{N}$ is denoted by $\mathfrak{M}\times\mathfrak{N}$ whereas $\mathfrak{M}^n$ denotes the Cartesian product of n lattices identical with $\mathfrak{M}$. The symbol $\mathfrak{M}+\mathfrak{N}$ denotes the sum of the lattices $\mathfrak{M}$ and $\mathfrak{N}$ obtained as follows: let $\mathfrak{M}$ be isomorphic to $\mathfrak{M}'$, $\mathfrak{N}$ to $\mathfrak{N}'$ and $\mathfrak{M}'\cap\mathfrak{N}' = = 1_{\mathfrak{M}'} = 0_{\mathfrak{N}'}$. Then the system $\mathfrak{M}'\cup\mathfrak{N}'$ as a partially ordered system is a lattice isomorphic to $\mathfrak{M}+\mathfrak{N}$ [4]. The symbol $n\mathfrak{M}$ denotes the result of summing up of the n lattices identical with $\mathfrak{M}$. The sum of lattices

---

[1] This is a summary of a report presented by the author on April 26, 1970 to the XVIth Conference of the History of Logic organized in Cracow by the Section of Logic, Polish Academy of Sciences, together with the Department of Logic of the Jagiellonian University.

[2] For terminology cf. [15].

[3] We use parenthesis-free notation of Łukasiewicz explained in [13].

[4] The sum $\mathfrak{M}+\mathfrak{N}$ is defined up to isomorphism (cf. [19]).

may be visualized by means of Hasse diagrams. For instance let the diagrams of the lattices $\mathfrak{M}$ and $\mathfrak{N}$ be as follows:

$$\mathfrak{M} = 2 \diamondsuit_4^1 3 \qquad \mathfrak{N} = \begin{array}{c} a \\ b \\ c \end{array}$$

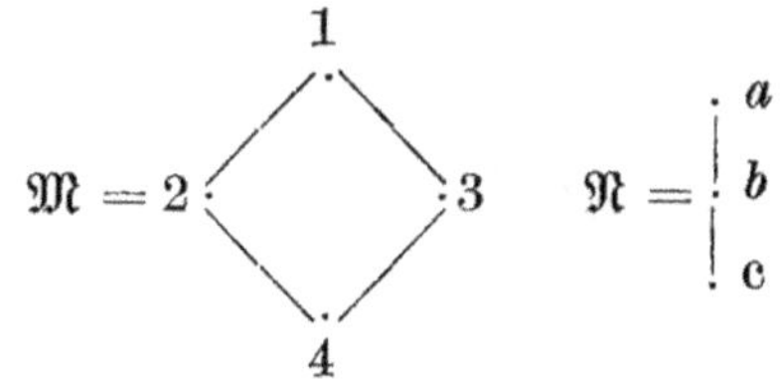

To obtain the sum $\mathfrak{M}+\mathfrak{N}$ we put

$$\mathfrak{M}' = \mathfrak{M} \quad \text{and} \quad \mathfrak{N}' = \begin{array}{c} a \\ b \\ 1 \end{array}$$

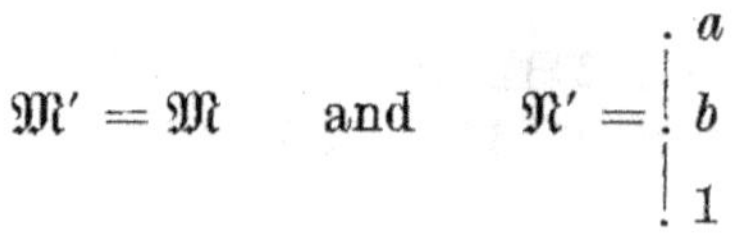

and then the system $\mathfrak{M}' \cup \mathfrak{N}'$ may be described by the following diagram:

$$\mathfrak{M}+\mathfrak{N} \cong 2 \diamondsuit_4^{\begin{array}{c} b \\ a \\ 1 \end{array}} 3$$

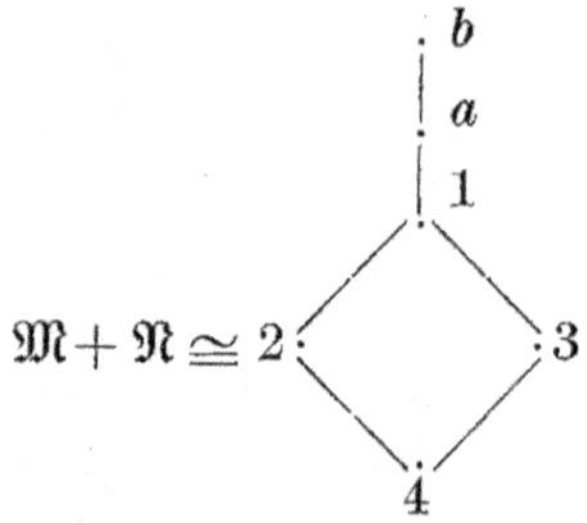

Thus, the diagram of the lattice $\mathfrak{M}+\mathfrak{N}$ may be constructed by placing the diagram of the lattice $\mathfrak{N}$ over the diagram of the lattice $\mathfrak{M}$. Both Cartesian product and the sum of the lattices may be extended to the case ω-infinite. So we have $\mathfrak{M}^\omega$ and $\omega\mathfrak{M}$. In terms of the Cartesian product and the sum the Gödel's lattices from [4] and Jaśkowski's lattices from [11] may be described in a very simple way. The sequence of Gödel's lattices $\{\mathfrak{G}_n\}$ is given by the condition $\mathfrak{G}_n = n\mathfrak{B}$ and the sequence of Jaśkowski's lattices $\{\mathfrak{J}_n\}$ by the condition $\mathfrak{J}_0 = \mathfrak{B}$, $\mathfrak{J}_{n+1} = \mathfrak{J}_n^{n+1} + \mathfrak{B}$. In Dummett and Lemmon [3] and also in Grzegorczyk [5] a still simpler description of Jaśkowski's lattices is to be found. This description is based on the observation that if t_1, t_2 are two disjoint trees with corresponding lattices of the open subsets $\mathfrak{T}_1$, $\mathfrak{T}_2$, then the lattice $\mathfrak{T}_1 \times \mathfrak{T}_2 + \mathfrak{B}$ is isomorphic to the lattice of the open subsets on the tree obtained from t_1 and t_2 by adding a new common top point, i.e., on the tree constructed as follows:

$$\overset{\displaystyle\cdot}{\diagup\ \diagdown}$$
$$t_1 \qquad\quad t_2$$

Since $\mathfrak{B}$ is the lattice of the open subsets on the one-point tree, then the successive Jaśkowski's lattices are the lattices of open subsets on the following trees:

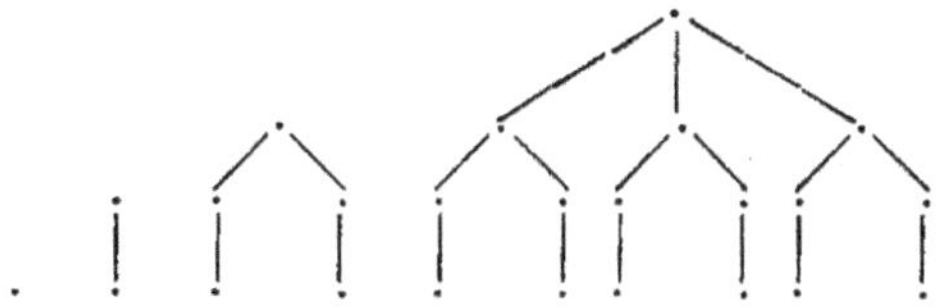
etc.

The history of the axiomatization of the lattice $\mathfrak{B}$ is well-known, but the fact that the lattice $2\mathfrak{B}$ was first axiomatized in 1938 by Łukasiewicz [12] seems to the author to be rather forgotten. The fact that the authors who repeat this result of Łukasiewicz in later years do not refer to him provides grounds for our opinion. The axiomatics given by Łukasiewicz consists of the Ax and the following axiom $\mathfrak{L} = INbaIIIabaa$. The lattice $\omega\mathfrak{B}$ was axiomatized in 1959 by Dummett [2]. Dummett's axiomatics contains Ax and the following axiom $D = AIabIba$. The whole sequence of Gödel's lattices was axiomatized in 1962 by Thomas [18]. Thomas defined the following sequence of axioms $T_1 = IIIa_1a_2a_1a_1$, $T_{n+1} = IIIa_{n+1}a_{n+2}a_1T_n$ and proved that $Cn(T_n) = E(n\mathfrak{B})$. The result of Thomas was repeated in 1966 by Hosoi [7] who does not refer to the forerunner. The series of results stated by Jankov in [8][5] deserve a more detailed treatment. Jankov takes into consideration the formulas $J_1 = ANNaNa$, $J_2 = IKKNNaIabIIbaab$[6] and the following kinds of lattices:

(i) sequence $\{\mathfrak{D}_n\}$ such that $\mathfrak{D}_n$ is the sublattice generated by the set of dense elements of the Jaśkowski's lattice $\mathfrak{J}_n$, where $\mathfrak{D}_n$ consists of all dense elements of $\mathfrak{J}_n$ and $0_{\mathfrak{J}_n}$,

(ii) sequence $\{\mathfrak{B}^n + \mathfrak{B}\}$.

Jankov states that:

(1) $Cn(J_1) = \bigcap_n E(\mathfrak{D}_n)$

(2) $Cn(J_2) = \bigcap_n E(\mathfrak{B}^n + \mathfrak{B})$

(3) $Cn(J_1, J_2) = E(2\mathfrak{B})$.

For every formula A the following conditions hold:

(4) $Cn(A) = E(\mathfrak{B})$ iff $A \in E(\mathfrak{B})$ and $A \notin E(2\mathfrak{B})$

(5) $Cn(A) = Cn(J_1)$ iff $A \in \bigcap_n E(\mathfrak{D}_n)$ and $A \notin E(\mathfrak{B}^2 + \mathfrak{B})$

---

[5] For the other interesting results of Jankov cf. [9] and [10].

[6] It is easy to prove that $Cn(J_1, J_2) = Cn(T_2)$ and $Cn(J_2) = Cn(INNa_2T_1) = Cn(IIIa_2a_3a_2T_1)$.

(6) $Cn(A) = Cn(J_2)$ iff $A \in \bigcap_n E(\mathfrak{B}^n + \mathfrak{B})$ and $A \notin E(3\mathfrak{B})$

(7) $Cn(A) = Cn(J_1, J_2)$ iff $A \in E(2\mathfrak{B})$, $A \notin E(3\mathfrak{B})$ and $A \notin E(\mathfrak{B}^2 + \mathfrak{B})$.

The items (4), ..., (7) give convenient criteria for determining whether an intermediate calculus is equivalent to one of the four above. The item (4) was repeated without referring to Jankov by Hanazawa in [6]. The item (7) was generalized in Anderson's [1]. Anderson proves that:

$$Cn(A) = Cn(T_n) \text{ if } A \in E(n\mathfrak{B}),\ A \notin E(n+1\mathfrak{B}),$$
$$A \notin E(\mathfrak{B}^2 + \mathfrak{B}) \text{ and } A \notin E(\mathfrak{B} + (\mathfrak{B}^2 + \mathfrak{B}))$$

which in case $n = 2$ is equivalent to item (7) [7]. Besides Anderson gives in [1] the Jankov-type criterion for the previously described calculus of Dummett proving that $Cn(A) = Cn(D)$, if and only if $A \in E(\omega\mathfrak{B})$, $A \notin E(\mathfrak{B}^2 + \mathfrak{B})$ and $A \notin E(\mathfrak{B} + (\mathfrak{B}^2 + \mathfrak{B}))$. In his [19] Troelstra gives information that de Jongh has achieved an (unpublished) result that every finite lattice is finitely axiomatizable. The very elegant proof of de Jongh's result is published by McKay in [14].

References

[1] Anderson J. G.: *An application of Kripke's completeness theorem for intuitionism to superconstructive propositional calculi.* Zeitschrift für mathematische Logik und Grundlagen der Mathematik, 15 (1969), 259–288.

[2] Dummett M.: *A propositional calculus with a denumerable matrix.* Journal of Symbolic Logic, 24 (1959), 79–106.

[3] Dummett M., Lemmon E. J.: *Modal logics between S. 4 and S. 5.* Zeitschrift für mathematische Logik und Grundlagen der Mathematik, 5 (1959), 250–264.

[4] Gödel K.: *Zum intuitionistischen Aussagenkalkül.* Akademie der Wissenschaften in Wien. Mathematisch-naturwissenschaftliche Klasse, Anzeiger, 69 (1932), 65–66.

[5] Grzegorczyk A.: *A philosophically plausible formal interpretation of intuitionistic logic.* Indagationes Mathematicae, 26 (1964), 596–601.

[6] Hanazawa M.: *A characterization of axiom schema playing the rôle of tertium non datur in intuitionistic logic.* Proceedings of the Japan Academy of Science, 42 (1966), 1007.

[7] Hosoi T.: *The separable axiomatization of the intermediate propositional systems S_n of Gödel.* Proceedings of the Japan Academy of Science, 42 (1966), 1001.

[8] Jankov V. A.: *On certain superconstructive propositional calculi.* Doklady Akademii Nauk SSSR, 151 (1963), 796–798.

[7] Because there is a sublattice of $(\mathfrak{L} + (\mathfrak{L}^2 + \mathfrak{L}))$ isomorphic to $3\mathfrak{L}$.

[9] JANKOV V. A.: *On the realizable formulae of propositional logic.*
Doklady Akademii Nauk SSSR, 151 (1963), 1035–1037.

[10] JANKOV V. A.: *On the relation between deducibility in intuitionistic propositional calculus and finite implicative structures.*
Doklady Akademii Nauk SSSR, 151 (1963), 1293–1294.

[11] JAŚKOWSKI S.: *Recherches sur le système de la logique intuitionniste.*
Actes du Congrès International de Philosophie Scientifique VI. Philosophie des Mathématiques.
Actualités Scientifiques et Industrielles, 393 (1936), 58–61.

[12] ŁUKASIEWICZ J.: *Die Logik und das Grundlagenproblem.*
In: Les entretiens de Zürich sur les fondements et la méthode des sciences mathématiques, 6–9 XII 1938.
Zürich 1941, 82–100.

[13] ŁUKASIEWICZ J.: *Elementy logiki matematycznej.*
Warszawa 1958.

[14] McKAY C. G.: *On finite logics.*
Indagationes Mathematicae, 29 (1967), 363–365.

[15] RASIOWA H., SIKORSKI R.: *The mathematics of metamathematics.*
Warszawa 1968.

[16] STONE M. H.: *Topological representation of distributive lattices and Brouwerian logics.*
Časopis pro pěstováni matematiky, 67 (1937), 1–25.

[17] TARSKI A.: *Der Aussagenkalkül und die Topologie.*
Fundamenta Mathematicae, 31 (1938), 103–134.

[18] THOMAS I.: *Finite limitations of Dummett's LC.*
Notre Dame Journal of Formal Logic, 3 (1962), 170–174.

[19] TROELSTRA A. S.: *On intermediate propositional logics.*
Indagationes Mathematicae, 27 (1965), 141–152.

JAN WOLEŃSKI

ON ACKERMANN'S RIGOROUS IMPLICATION [1]

1. Historical remarks

The strict implication introduced by C. I. Lewis preserves several non-intuitive properties. In this connection some other approaches towards implication were given, in order to fulfill some more intuitions connected with ordinary „if..., then...". One of such approaches is the concept of rigorous implication introduced by W. Ackermann. In [1] Ackermann gave motives which led him to the search for a stronger concept of implication than this of strict implication. He formed a system of rigorous implication by means of the axiomatic method and also as a calculus of sequents and then he extended these systems to modal calculi. Ackermann's calculi do not belong to these systems which are intensively studied in contemporary logic [2]. Their founder himself was examining relations between the strict and rigorous implication (cf. [2]). The systems of rigorous implication inspired the authors of the system E (cf. [3], [4], [5], [6]). Dončenko [7] was examining the decision problem for the axiomatic system of rigorous implication. Zinovev [18], [19] was concerned with rigorous implication from the point of view of the intuitions associated with the concept of entailment. The works of Maksimova [11], [12] also should be mentioned although they do not refer straight to Ackermann's systems but rather to the weaker ones because in case of the Ackermann's systems t is difficult to solve the decision problem. Finally, we should as well

---

[1] This paper is an extended version of a lecture given by the author on April 24, 1969 to the XVth Conference of the History of Logic organized in Cracow by the Section of Logic, Polish Academy of Sciences, together with the Department of Logic of the Jagiellonian University.

[2] Ackermann did not admit that his system has special importance for logic. He was claiming that the concept of rigorous implication is more interesting from the philosophical point of view. Cf. [9], 37.

mention the work of Pogorzelski [14] in which he deals with a very important problem of the deduction theorem for the propositional calculi.

2. The classical logical calculus

The starting point for Ackermann are two formalizations of the classical propositional logic.

First of them is the classical propositional calculus as the calculus of sequents (System Σ).

Let $P, Q, R, \ldots$ denote formulas of the propositional calculus. Let C denote implication, A — disjunction, K — conjunction and N — negation.

An expression having one of the following forms is called a sequent:

$$P \vdash Q$$

$$P, Q \vdash R$$

$$P \vdash$$

$$P, Q \vdash$$

$$\vdash P .$$

The expressions $P, Q \vdash R$, $P \vdash$, $\vdash P$ may be read as „R is deducible from P and Q", „every conclusion is deducible from P", „P is deducible from every premise".

The sequents having one of the following forms are basic.

$$\vdash CPP$$

$$KPQ \vdash P$$

$$KPQ \vdash Q$$

$$P \vdash APQ$$

$$Q \vdash APQ .$$

There are the following rules of inference.

(Ia). From the sequent with at least one premise and conclusion there follows the sequent which results from the first by prefixing an arbitrary premise and a conclusion by CS.

(Ib). From the sequent $P \vdash AQR$ there follows $CQS, CRS \vdash CPS$.

(II). From sequents $F, Q \vdash R$ and $\vdash Q$ there follows $F \vdash R$.

In (II) and in the next rules the symbol F may be replaced by one or two formulas or by the empty formula. Of course, the replacement must be made in such a way that the resulting expression is a sequent.

(III). From the sequent $F \vdash CQR$ there follows $F, Q \vdash R$.

(IV). From the sequent $P, Q \vdash F$ there follows $Q, P \vdash F$.

(V). From the sequent $KPQ \vdash F$ there follows $P, Q \vdash F$.

(VIa). From a given sequent there follows every sequent which results from the first by means of putting the conclusion prefixed by N in place of some premise and putting that premise prefixed by N in place of the conclusion (the empty formula may also be a premise or a conclusion).

(VIb). From a given sequent there follows every sequent which results from the first by means of putting NNP in place of P or P in place of NNP. The formula $P(NNP)$ is a premise of the first sequent.

The second formalization, i.e., the axiomatic system of the propositional calculus (System Π) is the system of Hilbert-Bernays [10]. It is obvious that the systems Π and Σ are equivalent, i.e., for every formula P, P is a thesis of the system Π if and only if the sequent $\vdash P$ is a thesis of the system Σ.

3. The systems of rigorous implication

The rigorous implication is supposed to be a formalization of the logical relation between two propositions which holds when the content of the consequent is contained in the content of antecedent. This relation does not depend on the logical values of the antecedent and the consequent but on the content relation.

The system of rigorous implication conceived of as a calculus of sequents (System Σ') may be presented as follows.

An expression having one of the following forms is called a sequent:

$$P \vdash Q$$

$$P, Q \vdash R$$

$$P^*, Q \vdash R$$

$$P^*, Q^* \vdash R$$

$$P \vdash$$

$$\vdash P$$

$$P, Q \vdash$$

$$P^*, Q \vdash$$

$$P^*, Q^* \vdash$$

The definitions below enable us to operate with starred premises:

$$P^*, Q \vdash R \overset{\text{df}}{=} P \vdash CQR$$

$$P^*, Q^* \vdash R \overset{\text{df}}{=} P \vdash CQR \quad \text{and} \quad Q \vdash CPR$$

$$P^*, Q \vdash \overset{\text{df}}{=} P \vdash NQ$$

$$P^*, Q^* \vdash \overset{\text{df}}{=} P \vdash NQ \quad \text{and} \quad Q \vdash NP\,.$$

The sequents having one of the following forms are basic:

$$\vdash CPP$$

$$KPQ \vdash P$$

$$KPQ \vdash Q$$

$$P \vdash APQ$$

$$Q \vdash APQ$$

$$P, AQR \vdash AQKPR$$

There are the following rules of inference.

(I′a). From the sequent with at least one premise and a conclusion there follows the sequent which results from the first by prefixing an arbitrary premise and the conclusion by CS in such a way that every non-starred premise of the first sequent is prefixed. If the first sequent has two premises, then the resulting sequent has every premise starred. However, in case when the first sequent has two non-starred premises, then both premises of the resulting sequent are non-starred.

(I′b). From the sequent $P \vdash AQR$ there follows $CQS, CRS \vdash CPS$ where CQS and CRS are non-starred.

(II′a). From sequents $P, Q \vdash R$, $\vdash P$ and $\vdash Q$ there follows $\vdash R$.

(II′b). From sequents $F, P \vdash Q$ and $\vdash P$ (where F is starred or the empty formula) there follows $F \vdash Q$.

(II′c). From sequents $P, Q \vdash$ and $\vdash P$ there follows $Q \vdash$.

In the rules (II′a), (II′b), (II′c) it does not matter whether premises P and Q are starred or not.

(III′). From the sequent $F \vdash CQR$ there follows $F, Q \vdash R$. In the second sequent F is starred or the empty formula and Q is non-starred.

(IV′). From the sequent with two starred or two non-starred premises there follows the sequent with its premises commutated.

(V′). From the sequent $KPQ \vdash F$ there follows $P, Q \vdash F$ where both premises in the resulting sequent are non-starred.

(VI'a). From any sequent except the one which has two non-starred premises there follows every sequent resulting from the first by means of putting the conclusion prefixed by N in place of some premise and putting that premise prefixed by N in place of the conclusion (the empty formula may also be a premise or a conclusion). If a starred premise occurs in the same place in the resulting sequent as in the first, it is starred in the resulting sequent, if and only if it occurs in the first place. The premise of the resulting sequent which was obtained by means of prefixing the conclusion of the first sequent by N is never starred.

(VI'b). From a given sequent there follows every sequent which results by means of putting NNP in place of P or P in place of NNP. The formula $P(NNP)$ is a premise of the first sequent. This replacement preserves the starring of premises.

One should notice that as a result of applying the above mentioned rules it is not possible to obtain the expression in the form of $P, Q^* \vdash R$ which is not the sequent from the basic sequent.

The system of rigorous implication conceived of as an axiomatic system (System Π') may be presented as follows.

Formulas having one of the following are axioms:

(A1) CPP

(A2) $CCPQCCQRCPR$

(A3) $CCPQCCRPCRQ$

(A4) $CCPCPQCPQ$

(A5) $CKPQP$

(A6) $CKPQQ$

(A7) $CKCPQCPRCPKQR$

(A8) $CPAPQ$

(A9) $CQAPQ$

(A10) $CKCPRCQRCAPQR$

(A11) $CKPAQRAQKPR$

(A12) $CCPQCNQNP$

(A13) $CKPNQNCPQ$

(A14) $CPNNP$

(A15) $CNNPP$

There are the following rules of inference:

(R1) $\dfrac{P,\ CPQ}{Q}$

(R2) $\dfrac{P, Q}{KPQ}$

(R3) $\dfrac{P,\ ANPQ}{Q}$

(R4) $\dfrac{CPCQR, Q}{CPR}$ (where Q is a theorem of Π')

The following lemmas hold for the system Σ'.

LEMMA 1. *If* $F, P \vdash Q$ *(where* F *is starred or the empty formula), then* $F \vdash CPQ$ *is provable.*

LEMMA 2. $P \vdash P$ *is provable.*

LEMMA 3. *If* $P, Q \vdash R$ *is provable, then* $\vdash CKPQR$ *is provable.*
The following derived rules are admissible for the system Π'.

(DR1) $\dfrac{CPQ,\ CQR}{CPR}$

(DR2) $\dfrac{CPQ}{CCRPCRQ}$

(DR3) $\dfrac{CPQ}{CCQRCPR}$

(DR4) $\dfrac{CPCQR,\ CRS}{CPCQS}$

(DR5) $\dfrac{CKPQR}{CKCSPCSQCSR}$

(DR6) $\dfrac{CPCQR}{CPCCSQCSR \text{ and } CCSQCPCSR}$

(DR7) $\dfrac{CPCQR}{CCSPCCSQCSR \text{ and } CCSQCCSPCSR}$

By means of the above lemmas and derived rules it is not difficult to check that there holds the following

THEOREM 1. *For every formula P, P is thesis of the system Π' if and only if the sequent $\vdash P$ is thesis of the system Σ'.*

It is easy to see that both systems Π' and Σ' developed above are the modifications of Π and Σ. The reason for introducing the starred premises in Σ' is to eliminate the so-called „paradoxes of material implication", e.g., $CPCQP$, $CQCPP$, $CKPNPQ$.

Weakened version of the deduction theorem given by Pogorzelski expresses in the simple way some interesting properties of rigorous implication. Pogorzelski says referring to [6] that in the system having as the axioms

$$CCPQCCQRCPR$$

$$CCCPPQQ$$

$$CCPCPQCPQ$$

and the rule (R1) exactly the same theorems are provable as those provable by means of the axioms (A1), ..., (A4) and the rules R1 and (R4).

Let Ax denote the set of all formulas having one of the following forms:

$$CCPQCCQRCPR,\ CCCPPQQ,\ CCPCPQCPQ$$

and let $Cn(X)$ denote the smallest set containing X and closed under detachment. The Pogorzelski's deduction theorem may be expressed by means of these conventions as follows:

If $Q \in Cn(Ax \cup \{P\})$, then $Q \in Cn(Ax)$ or $CPQ \in Cn(Ax)$.

Thus, the implication CPQ holds, if P is necessary for deduction of Q from the axioms.

4. The rigorous implication and modality

Both systems of the rigorous implication may be extended to the system with modalities. To define the modal concepts Ackermann introduces a new symbol f (absurdity).

The following rule is added to Σ'.

(VII′). From the sequent of the form of $F \vdash f$ there follows the sequent $F \vdash$ and conversely (F can not be the empty formula). This rule used in the first direction preserves the starring of premises while in the second it does not.

The axiomatic system with modalities (System Π'') equivalent to Σ' with the added rule (VII′) is obtained by adding to Π' the following axioms:

(A16) $CCPfNP$

(A17) $CKPNPf$

and the rule:

(R5) $$\frac{CPQ,\ CKCPQRf}{CRf}$$

Ackermann proves that rules (R3) and (R4) are dispensable in Π''.

By adding to Π'' the following definitions of necessity (L), impossibility (U) and possibility (M) we get a modal logic.

$$LP \overset{\mathrm{df}}{=} CNPf$$

$$UP \overset{\mathrm{df}}{=} CPf$$

$$MP \overset{\mathrm{df}}{=} NCPf$$

In Π'' some formulas which are the paradoxes of strict implication are improvable, e.g., $CLPCQP$, $CUPCNPQ$.

Ackermann's way of defining modalities in Π'' was discussed by Robinson [16], Anderson and Belnap [4]. Anderson and Belnap point out that necessity can be defined by means of the method due to Prior [15], e.g.,

$$LP \overset{\mathrm{df}}{=} CCPPP$$

already in Π' and resulting systems is equivalent to Π''.

Ackermann [2] states that the system $S2$ can be interpreted in Π''. The proof may be sketched as follows.

Let us consider the system $S2$ formalized as in Schmidt [17]. It may be proved that Schmidt's axioms and rules for $S2$ are provable in Π'', if the implication symbol occurring in them is replaced by the symbol C_1 defined as follows:

$$C_1PQ \overset{\mathrm{df}}{=} CKKPNQNff\,.$$

To prove this it is convenient to have

LEMMA 4. *If CPQ is provable in Π'', then C_1PQ is provable.*

Thus, C_1 has all properties of strict implication. However, to prove the converse, e.g., that strict implication has all properties of C_1, the decision problem for Π'' should be solved.

5. Some properties of Π' and Π''

At this point we shall present some theorems concerning Π' and Π''. The two of them were given by Ackermann himself.

THEOREM 2. *Every tautology of the classical propositional calculus built up exclusively by means of N, K and A is provable in Π'.*

Proving this Ackermann makes use of the fact that every tautology of the classical propositional calculus in the conjunctive normal form is provable in Π'.

Using Theorem 2 Mleziva [13] interpretes in the trivial way the classical propositional calculus in Π'.

Ackermann gives the tables which enable him to prove the improvability of some formulas in Π''.

C	1	2	3	4	5	6	N
1	3	6	6	6	6	6	6
2	3	3	6	6	6	6	5
3	3	3	3	6	6	6	4
4	3	6	6	3	6	6	3
5	3	3	6	3	3	6	2
6	3	3	3	3	3	3	1

K	1	2	3	4	5	6
1	1	2	3	4	5	6
2	2	2	3	5	5	6
3	3	3	3	6	6	6
4	4	5	6	4	5	6
5	5	5	6	5	5	6
6	6	6	6	6	6	6

A	1	2	3	4	5	6
1	1	1	1	1	1	1
2	1	2	2	1	2	2
3	1	2	3	1	2	3
4	1	1	1	4	4	4
5	1	2	2	4	5	5
6	1	2	3	4	5	6

Reducing the given formula we replace any variable by 5 and f by 4. Now it is easy to show that every provable formula will be assigned the value 1, 2 or 3.

THEOREM 3. *The formula of the type $CPCQR$ is not provable in Π'', if in P there is no C or f.*

Very interesting theorems about Π' may be found in Dončenko [7] who states

THEOREM 4. *The formula CPQ is not provable in Π', if P and Q do not have at least one variable in common.*

Proving this theorem one needs the matrix

$$\mathfrak{M} = \langle\{1, 2, \dots, 8\}, \{1, 2, \dots, 4\}, C, N, K, A\rangle .$$

In that matrix

$$Kxy = \min(x, y)$$

$$Axy = \max(x, y)$$

and operations C and N are given by the following table

C	1	2	3	4	5	6	7	8	N
1	8	8	8	8	8	8	8	8	8
2	1	2	8	8	6	6	8	8	6
3	1	8	3	8	7	8	7	8	7
4	1	2	3	4	5	6	7	8	5
5	1	8	8	8	4	8	8	8	2
6	1	2	8	8	2	2	8	8	2
7	1	2	3	8	3	8	3	8	3
8	1	1	1	1	1	1	1	1	1

It is easy to see that all theorems of Π' are satisfied in $\mathfrak{M}$, i.e., they have one of the values 1, 2, 3, 4 with respect to any valuation of the variables. The following simple lemma can be proved by induction.

LEMMA 5. *If v is a valuation such that every variable has the value 2 (every variable has the value 3), then every formula P has the value 2 or 6 (3 or 7) with respect to valuation v.*

Let us assume that P and Q are formulas which have no variable in common. Let v be such a valuation that the variables of Q have the value 3 with respect to v. By means of lemma 5 we get that $v(P) = 2$ or $v(P) = 6$ and $v(Q) = 3$ or $v(Q) = 7$. Now, considering all the possible cases we have $v(CPQ) = 8$, thus CPQ is not provable in Π'.

Dončenko points out that the matrix $\mathfrak{M}$ used in the proof of the theorem 4 is not a characteristic matrix for Π'. For instance, the formula $CCPCQRCQCCPR$ is satisfied in $\mathfrak{M}$ and it is improvable in Π'.

THEOREM 5. *There does not exist a finite matrix characteristic for Π'.*

To prove this theorem let us define the sequence of matrices

$$\mathfrak{N}_n = \langle U_n, W_n, C_n, K_n, A_n, N_n\rangle$$

where

$$U_n = \{-n, \ldots, -1, 1, \ldots, n\}$$

$$W_n = \{1, \ldots, n\}$$

$$C_n xy = \begin{cases} \max(|x|, |y|) & \text{for} \quad x \leqslant y \\ \min(|-x|, |-y|) & \text{for} \quad x > y \end{cases}$$

$$K_n xy = \min(x, y)$$

$$A_n xy = \max(x, y)$$

$$N_n = -x$$

It may be shown that for every n, the matrix $\mathfrak{N}_n$ satisfied all theorems of Π'.

Let us assume that the matrix $\mathfrak{A}$ is characteristic for Π' and has k elements. Now we consider the formula S which is the disjunction of all formulas CP_iP_j where $i = 1, \ldots, k$, $j = 2, \ldots, k+1$ and $i < j$. As the matrix $\mathfrak{A}$ has k elements it is evident that for every valuation at least two variables from $P_1, \ldots, P_{k+1}$ have the same value. Thus, at least one formula CP_iP_j is satisfied. Now, by means of (A8) we have that S is also satisfied. Hence S is provable in Π', if $\mathfrak{A}$ is characteristic. On the other hand, the valuation v in $\mathfrak{N}_{k+1}$ such that for any $i = 1, \ldots, k+1$, $v(P) = -i$ gives the value -2 to the formula S and we have a contradiction. This completes the proof of theorem 5.

It seems to be interesting to examine whether there exists a sequence of finite matrices characteristic for the system Π', e.g., whether this system has the so-called finite model property in the sense of Harrop [8] [3].

References

[1] ACKERMANN W.: *Begründung einer strenger Implikation.* Journal of Symbolic Logic, 21(1956), 113–128.

[2] ACKERMANN W.: *Über die Beziehung zwischen strikten und strengen Implikation.* Dialectica, 12(1958), 213–222.

[3] ANDERSON A. R.: *Some open problems concerning the system E of Entailment.* Proceedings of a Colloquium on Modal and Many-Valued Logic. Helsinki 1963, 7–18.

[4] ANDERSON A. R., BELNAP N. D.: *Modalities in Ackermann's „rigorous implication".* Journal of Symbolic Logic, 24(1959), 107–111.

[5] ANDERSON A. R., BELNAP N. D.: *The pure calculus of entailment.* Journal of Symbolic Logic, 27(1962), 19–52.

[6] ANDERSON A. R., BELNAP N. D., WALLACE J. R.: *Independent axiom schemata for the pure theory of entailment.* Zeitschrift für mathematische Logik und Grundlagen der Mathematik, 6(**1960**), 93–95.

[7] DONČENKO V. V.: *Nekotorye woprosy swjazannye s problemoj razrešenija dlja isčislenija strogoj implikacii Akkermana.* In: Problemy logiki. Moskva 1963, 18–24.

[8] HARROP R.: *On the existence of finite models and decision procedures for propositional calculi.* Proceedings of the Cambridge Philosophical Society, 54(1958), 1–13.

[9] HILBERT D., ACKERMANN W.: *Grundzüge der mathematischen Logik.* Berlin 1959.

[3] This problem was raised by dr A. Wroński. The author is indebted to him for valuable remarks on the present paper.

[10] Hilbert D., Bernays P.: *Grundlagen der Mathematik*. Vol. I. Berlin 1934.

[11] Maksimova L. L.: *Formal'nye wywody w isčislenii strogoj implikacii.* Algebra i Logika, 5(1966), 33–39.

[12] Maksimova L. L.: *Ob isčislenii strogoj implikacii.* Algebra i Logika, 7(1968), 57–76.

[13] Mleziva M.: *Über das Enthaltensein des klassischen Aussagenkalkül in den nicht-klassischen Aussagenkalkülen.* Praha 1966.

[14] Pogorzelski W. A.: *Przegląd twierdzeń o dedukcji dla rachunków zdań.* Studia Logica, 15(1964), 163–178.

[15] Prior A. N., Lemmon E. J., Meredith C. A., Meredith D., Thomas I.: *Calculi of pure strict implication.* Canterbury 1957.

[16] Robinson A.: *Review of Ackermann* [1]. Mathematical Review, 28(1957), 117.

[17] Schmidt A.: *Ein rein aussagenlogischer Zugang zu den Modalitäten der strikten Logik.* Proceedings of the International Mathematical Congress, (2) 1954, 407–408, Amsterdam.

[18] Zinovev A. A.: *Logika wyskazywanij i teorija wywoda.* Moskva 1962.

[19] Zinovev A. A.: *Logičeskoe sledovanie.* Problemy logiki i teorii poznanija. Moskva 1968.

II. CONTRIBUTIONS TO THE HISTORY OF COMPLETENESS OF THE FIRST ORDER PREDICATE CALCULUS

JAN ZYGMUNT

KURT GÖDEL'S DOCTORAL DISSERTATION [1]

1. Introduction

Kurt Gödel, one of the most remarkable contemporary mathematicians and logicians was born on April 28, 1906 at Brünn (Brno). He studied mathematics and physics at the University of Vienna and took the Ph. D. degree there after presenting his work entitled *Die Vollständigkeit der Axiome des logischen Funktionenkalküls* [2]. The degree was granted on 6 February 1930.

The dissertation presents for the first time in the history the proof of the completeness theorem for the first order predicate calculus described in the first volume of *Principia Mathematica* (published in 1910). Besides, there is also presented the compactness theorem and the results achieved in this field are applied to the first order predicate calculus with identity. In the final part of the dissertation Gödel is interested in the problem of independence of the axioms of the first order predicate calculus with identity. Thus Gödel's dissertation is one of the great works about the predicate calculus described in *Principia Mathematica* and the basic pioneer work in the theory of models. The author of the dissertation was familiar with the results which were described in Hilbert and Ackermann's monograph [3], Skolem's and Löwenheim's papers [7], [5] and in Bernays' [1]. Let us notice, by the way, that Gödel does not cite Post's paper [6] in which there was presented the completeness theorem for the classical propositional calculus of *Principia Mathematica.*

---

[1] This paper is an extended version of a lecture delivered by the author on April 25, 1970 to the XVIth Conference of the History of Logic organized in Cracow by the Section of Logic, Polish Academy of Sciences, together with the Department of Logic of the Jagiellonian University.

2. The completeness theorem and the compactness theorem for the first order predicate calculus

While presenting Gödel's results we try to reproduce exactly the way of his argumentation. Gödel uses the symbolism of Hilbert and Ackermann's monograph [3].

The alphabet of the first order predicate calculus is as follows: a denumerable set $\{x, y, z, \ldots\}$ of symbols called individual variables; a non--empty set $\{F, G, H, \ldots\}$ of symbols called predicate variables: a denumerable set $\{X, Y, Z, \ldots\}$ of symbols called propositional variables; $\vee$, $\overline{}$, (x) denote logical constants, i.e., disjunction, negation and the universal quantifier binding the variable x; by means of familiar abbreviations the following logical constants are defined: $\rightarrow$ (implication), & (conjunction), $\sim$ (equivalence), (Ex) (the existential quantifier binding the variable x).

In the present paper metamathematical symbols: $x_0, x_1, x_2, \ldots$ are used for individual variables; $A, B, C, \ldots$ for any formulas built up in the discussed alphabet. (P), (Q), (R) with natural indices at the bottom or without, denote the sequences of symbols whose terms are the quantified variables, e.g., $(P_2) = (x)(Ey)(z)$. The letters α, β, γ, δ with natural indices at the bottom or without, denote finite sequences of the individual variables. If, e.g., $\alpha = \langle x, y, \ldots, z\rangle$, then (α) and $(E\alpha)$ are respectively the abbreviations of the symbols $(x)(y)\ldots(z)$ and $(Ex)(Ey)\ldots(Ez)$, whereas if any variable in the sequence α occurs more than once then in (α) and in $(E\alpha)$ it occurs only once.

Gödel's investigations are concerned with the predicate calculus based upon the following axioms:

(A1) $X \vee X \rightarrow X$

(A2) $X \rightarrow X \vee Y$

(A3) $X \vee Y \rightarrow Y \vee X$

(A4) $(X \rightarrow Y) \rightarrow [(Z \vee X) \rightarrow (Z \vee Y)]$

(A5) $(x)F(x) \rightarrow F(y)$

(A6) $(x)[X \vee F(x)] \rightarrow [X \vee (x)F(x)]$
where the formula $X \rightarrow Y$ is an abbreviation of the formula $\bar{X} \vee Y$.

The rules of deduction are as follows:

(R1) the rule of detachment,

(R2) the rule of substitution for the propositional and predicate variables,

(R3) the rule of generalization: from $A(x)$, $(x)A(x)$ may be inferred,

(R4) individual variables (free or bound) may be replaced by any others, so long as this does not cause overlapping of the scopes of variables denoted by the same sign.

The above formalization of the predicate calculus is given in *Principia Mathematica* vol. 1, *1 and *10. Gödel omits the axiom of associativity for the disjunction which depends on axioms (A1)–(A4), as it was proved by Bernays in [1] and which occurs in *Principia Mathematica.* The authors of *Principia Mathematica* use the rules (R1)–(R4) although they do not formulate all of them explicitly.

We read the symbol $S \vdash A$ as follows: A is a consequence of the set S of formulas; a symbol $\vdash A$ is to be read: A is a thesis of the discussed calculus. By $M = \langle U, f, g, h, \ldots, w_1, w_2, w_3, \ldots\rangle$ we shall denote a pseudo-model of the discussed language. In the sequence M the set U is non-empty; $f, g, h, \ldots$ are relations defined in the domain U and corresponding to the symbols $F, G, H, \ldots$, while $w_1, w_2, w_3, \ldots$ are logical values t (truth) or f (falsity) which correspond to the propositional variables $X, Y, Z, \ldots$ If $A(x_0, \ldots, x_n)$ is a formula in which the only free variables are $x_0, \ldots, x_n$, then the mark

$$M \models A(x_0, \ldots, x_n)[t_0, \ldots, t_n]$$

means that the sequence of elements $t_0, \ldots, t_n$ which belong to U, satisfies the formula A. The formula A is called satisfiable in M, if there is a sequence of elements from the universe U, which satisfies it. A is called valid in M if A is satisfiable by every sequence of elements from the universe of M. A is valid, if it is valid in every M. A is satisfiable, if there is a model M such that A is satisfiable in M. A set S of the formulas is called (simultaneously) satisfiable, if there is a model M such that every formula which belongs to S is satisfiable in M.

The main theorem in the dissertation is

THEOREM 1. (the completeness theorem for the first order predicate calculus). *Every valid formula is a thesis.*

A formula has a property (Γ), if and only if it is satisfiable or it is a counter-thesis (that is its negation is a thesis).

THEOREM 2. *Every formula has the property* (Γ).

Theorem 2 is equivalent to theorem 1.

The series of lemmas below will allow us to carry on, in a concise way, the proof of theorem 1. The proofs of some of these lemmas can be found in Hilbert and Ackermann (cf. [3]).

Let us denote by a any sequence of the variables.

LEMMA 1. (a) $\vdash (a)F(a) \to (Ea)F(a)$

(b) $\vdash (a)F(a) \,\&\, (Ea)G(a) \to (Ea)[F(a) \,\&\, G(a)]$

(c) $\vdash (a)\overline{F(a)} \sim \overline{(Ea)F(a)}$.

LEMMA 2. *If* a_1 *is a sequence which is different from* a *only in the ordering of the variables, then* $\vdash (Ea)F(a) \to (Ea_1)F(a)$.

LEMMA 3. *If* a *is a sequence of distinct variables and the number of terms in the sequence* a_1 *is equal to the number of terms of* a, *then* $\vdash (a)F(a) \to (a_1)F(a_1)$.

LEMMA 4. *If*

(1) (p_i) *means one of the prefixes* (x_i) *or* (Ex_i), *where* $1 \leqslant i \leqslant n$,

(2) (q_i) *means one of the prefixes* (y_i) *or* (Ey_i), *where* $1 \leqslant i \leqslant m$,

(3) *every* x_i *is distinct from every* y_i,

(4) *a string of quantifiers* (P) *is formed from the* (p_i) *and* (q_i) *and satisfies the conditions*:

(i) *if* $1 \leqslant k < l \leqslant n$, *then prefix* (p_k) *precedes* (p_l),

(ii) *if* $1 \leqslant k < l \leqslant m$, *then prefix* (q_k) *precedes* (q_l),

then

$$\vdash (p_1)\ldots(p_n)F(x_1, \ldots, x_n) \,\&\, (q_1)\ldots(q_m)G(y_1, \ldots, y_m) \sim$$
$$\sim (P)[F(x_1, \ldots, x_n) \,\&\, G(y_1, \ldots, y_m)]\,.$$

LEMMA 5. *For every formula* A *there is a formula* N *in the normal form, such that* $\vdash A \sim N$. (Hilbert and Ackermann [3], Chap. 3, § 8).

LEMMA 6.

If

(1) $\vdash A(a) \sim B(a)$,
where a *consists entirely of distinct variables which are free both in* A *and* B,

(2) *for every* n *and for every* $i \leqslant n$
a sequence a_i *has the same number of terms as* a,

(3) $A(a_i)$ *and* $B(a_i)$ *are obtained from* $A(a)$ *and* $B(a)$ *by substitution*: a/a_i,

(4) *for every formula* A^* *there is* $i \leqslant n$ *such that*
$A(a_i)$ *is subformula of* A^* *and* B^* *is obtained from* A^* *by replacement of* $A(a_i)$ *by* $B(a_i)$

then

$\vdash A^* \sim B^*$ (Hilbert and Ackermann [3], Chap. 3, § 7).

LEMMA 7. *Every valid formula of the propositional calculus is a thesis, i.e., the axioms* (A1)–(A4) *together with the rules* (R1), (R2) *form a complete system of the propositional calculus.*

Let us denote by $\mathfrak{K}$ the set of all sentences in the normal form in which the string of all quantifiers has the shape: $(x)\ldots(Ey)$. It begins with a universal quantifier and it ends with an existential quantifier.

If $A \in \mathfrak{K}$, then the number k is called the degree of the formula A, if and only if k is the number of the quantifier strings of the following form: $(x_i)\ldots(x_j)(Ex_l)\ldots(Ex_n)$ which occur in the quantifier string in the formula A. The degree of A we denote by dgA.

Let us assume that $(P)A = (\alpha)(E\beta)A \in \mathfrak{K}$, $dg(P)A = 1$ and that the lengths of sequences α and β are r and s respectively.

Let us denote by $\{\alpha_i: i = 1, 2, 3, \ldots\}$ a sequence of the finite sequences chosen from the sequence of all variables $\langle x_0, x_1, x_2, \ldots\rangle$ and arranged according to the increasing sums of indices and in case when the sums of indices are equal we accept anty-lexicographical order determined by the sequence $\langle x_0, x_1, \ldots\rangle$. Namely:

$$\alpha_1 = \langle x_0, x_0, \ldots, x_0\rangle$$
$$\alpha_2 = \langle x_1, x_0, \ldots, x_0\rangle$$
$$\alpha_3 = \underbrace{\langle x_0, x_1, \ldots, x_0\rangle}_{r\text{-times}} \qquad \text{etc.}$$

Let us denote by $\{\beta_i: i = 1, 2, 3, \ldots\}$ a sequence of the finite sequences of the length s such that

$$\beta_n = \langle x_{(n-1)\cdot s+1}, \ldots, x_{n\cdot s}\rangle .$$

$\{A_i: i = 1, 2, 3, \ldots\}$ is a sequence of the formulas determined by the conditions:

$$A_1 = A(\alpha_1, \beta_1)$$
$$A_n = A(\alpha_n, \beta_n) \,\&\, A_{n-1} .$$

Let us accept the notation: $(P_n)A_n = (Ex_0)(Ex_1)\ldots(Ex_{n\cdot s})A_n$. It is easy to observe that all variables $x_0, \ldots, x_{n\cdot s}$ occur in A_n. So in $(P_n)A_n$ they are bound. The variables of the sequence α_{n+1} occur in P_n and they are different from all the variables of the sequence β_{n+1}, because $\beta_{n+1} = \langle x_{n\cdot s+1}, x_{n\cdot s+2}, \ldots, x_{(n+1)\cdot s}\rangle$. Let us denote by (P'_n) the prefix resulting from (P_n) by deletion of its all the quantifiers binding the variables of the sequence α_{n+1}. Without considering the order we have $(E\alpha_{n+1})(P'_n) = (P_n)$. With accepted notations the following holds:

LEMMA 8. *For every natural* n, $\vdash (P)A \to (P_n)A_n$.

The easy proof of this lemma in view of the lemmas above, we omit.

LEMMA 9. *Every formula of the first degree has the property* (Γ).

Proof. The formula $(P)A = (a)(E\beta)A$ of the first degree discussed in lemma 8 is built up of a number of predicate symbols and the propositional variables. Let $F_1, \ldots, F_k$ and $X_1, \ldots, X_l$ be all the mentioned symbols. F_i is an m_i — argument symbol. For every n, A_n is built up of the atoms of the shape

$$(*) \qquad F_1(x_{p_1}, \ldots, x_{p_{m_1}}), \ldots, F_k(x_{p_1}, \ldots, x_{p_{m_k}}).$$

Let us assign to every formula A_n, the formula Z_n of the propositional calculus, resulting from A_n by replacing in it the atoms $(*)$ by the propositional variables, whereas we assign to different atoms the distinct propositional variables (and distinct from $X_1, \ldots, X_l$). The sequence

$$M^{(n)} = \langle \{0, 1, 2, \ldots, n\cdot s\}, f_1^{(n)}, \ldots, f_k^{(n)}, w_1^{(n)}, \ldots, w_l^{(n)} \rangle$$

is called a model of level n of the formula $(P)A$, if and only if $f_i^{(n)}$ is an m_i — argument relation defined in the set $\{0, 1, 2, \ldots, n\cdot s\}$ which corresponds to the symbol F_i, while $w_i^{(n)}$ is a logical value t or f which correspond to the variable X_i and

$$M^{(n)} \models A_n(x_0, \ldots, x_{n\cdot s})[0, 1, \ldots, n\cdot s].$$

$M^{(n)}$ is a model of level n, if and only if Z_n is satisfiable, i.e., it takes the value t after assigning logical values, t or f, to all propositional variables occurring in Z_n. In view of lemma 7 every Z_n is either a counter-thesis or it is satisfiable. Let us first assume that there is n such that Z_n is a counter-thesis. Then on the basis of the rules (R2) and (R3) and lemma 1 (c) we conclude that $(P_n)A_n$ is a counter-thesis. From lemma 9 it results that $(P)A$ is a counter-thesis, i.e., $\vdash \overline{(P)A}$. Let us then assume that there is no n such that Z_n is a counter-thesis, i.e., for every n, Z_n is satisfiable. Then for every n there is a model $M^{(n)}$ of level n of the formula $(P)A$. Because the universes of these models are finite, so the number of the models of every level is finite. Considering that A_n has been constructed by successive conjunctions, it follows that there is an infinite sequence $\{M^{(i)}\}$ such that for every n, $M^{(n)} \subset M^{(n+1)}$. The last inclusion can be understood in the following way:

(i) the domain of $f_i^{(n)}$ is contained in the domain of $f_i^{(n+1)}$, in symbols, $D(f_i^{(n)}) \subset D(f_i^{(n+1)})$,

(ii) $f_i^{(n)} = f_i^{(n+1)} \restriction D(f_i^{(n)})$,

(iii) $w_i^{(n)} = w_i^{(n+1)}$.

The model M is defined as follows: $M = \langle U, f_1, \ldots, f_k, w_1, \ldots, w_l \rangle$ and

(i) $U = \{0, 1, 2, \ldots\}$,

(ii) $f_i(n_1, \ldots, n_{m_i})$ holds in M if and only if there is n such that $f_i^{(n)}(n_1, \ldots, n_{m_i})$ holds in $M^{(n)}$,

(iii) $w_i = w_i^{(n)}$, for every n.

The model M verifies the discussed formula $(a)(E\beta)A$. This completes the proof of lemma 9.

LEMMA 10. *If every formula of the degree k has a property (Γ), then every formula of degree $k+1$ has this property.*

Proof. Let us assume that for every formula B, if $dgB = k$, then B has the property (Γ). Besides, let $(P)A \in \mathfrak{R}$ and $dg(P)A = k+1$. From the definition of the set $\mathfrak{R}$ it follows that $(P) = (a)(E\beta)(Q)$, for some (Q), and $dg(Q)A = k$, where $(Q) = (\gamma)(E\delta)(R)$ for some γ, δ, and R (R may be empty symbol). Let F be a predicate symbol which does not occur in A. Let us introduce the two formulas

$$B = (a_1)(E\beta_1)F(a_1, \beta_1) \& (a)(\beta)[F(a, \beta) \rightarrow (Q)A]$$

$$C = (a_1)(a)(\beta)(\gamma)(E\beta_1)(E\delta)(R)\{F(a_1, \beta_1) \& [F(a, \beta) \rightarrow A]\},$$

where variable-sequences a, a_1, β, β_1, γ, δ are assumed to be pairwise disjoint. After easy transformations based upon lemmas 4 and 6 we get that $\vdash B \sim C$. Because $dgC = k$, then C has the property (Γ). Let us also observe that $B \rightarrow (P)A$ is a valid formula. If C is satisfiable then $(P)A$ also is satisfiable. If, on the other hand, $\vdash \overline{C}$, then $\vdash \overline{B}$. Putting in the place of F in the expression B, the formula $(Q)A$, we get that

$$\vdash \overline{(a_1)(E\beta_1)(Q)A \& (a)(\beta)[(Q)A \rightarrow (Q)A]}$$

and hence that $\vdash \overline{(a)(E\beta)(Q)A}$. In view of lemma 2 and lemma 3 we get that $\vdash \overline{(P)A}$. In both cases $(P)A$ has the property (Γ). The proof of the lemma 10 has been finished with the same.

THEOREM 3. *Every formula which belongs to the set $\mathfrak{R}$ has the property (Γ).*

Theorem 3 is a conclusion from lemmas 9 and 10.

THEOREM 4. *If every element which belongs to $\mathfrak{K}$ has the property (Γ), then every formula has the property (Γ).*

Proof. Let us assume that for any formula B if $B \in \mathfrak{K}$, then B has the property (Γ) and besides $A \notin \mathfrak{K}$. Let us denote by a the sequence whose terms are all free variables of A. Let us observe that $(Ea)A(a)$ is satisfiable, if and only if $A(a)$ is satisfiable and that $A(a)$ is a counter--thesis, if and only if $(Ea)A(a)$ is a counter-thesis. Let $(P)N$ be the normal form of the formula $(Ea)A(a)$. Let

$$B = (x)(P)(Ey)\{N \,\&\, [F(x) \vee \overline{F(x)}]\}\,,$$

where the x and y do not occur in (P). It is obvious that $B \in \mathfrak{K}$ and $\vdash B \sim (P)N$. The last equivalence is proved by means of lemma 4. From the above it follows that $(P)N$ has the property (Γ) and hence that A has the property (Γ).

From theorem 3 and theorem 4 there follows theorem 1.

Let us pass now to a generalization of theorem 1.

THEOREM 5. *Every denumerably infinite set S of the formulas is simultaneously satisfiable or there is a finite subset $S_f \subset S$ such that S_f is inconsistent.*

Theorem 5 follows from

THEOREM 6 (compactness theorem for the first order logic). *A denumerably infinite set S of the formulas is satisfiable, if and only if its every finite subset S_f is satisfiable.*

Proof. In view of theorem 4 and lemma 10 it is enough in the proof to limit oneself to the formulas of the first degree. Let

$$S = \{(a_1)(E\beta_1)A_1(a_1, \beta_1), \ldots, (a_n)(E\beta_n)A_n(a_n, \beta_n), \ldots\}$$

where a_i and β_i are the sequences of the individual variables of the length r_i and s_i respectively.

$\{a_k^i\colon k = 1, 2, \ldots\}$ is a sequence of all the sequences of the length r_i taken from the sequence $\langle x_0, x_1, \ldots\rangle$ and ordered according to the increasing sums of indices and for equal sums according to anty-lexicographical order determined by the sequence $\langle x_0, x_1, \ldots\rangle$.

$\{\beta_k^i\colon k = 1, 2, \ldots\}$ is a sequence whose terms are the sequences of the length s_i such that $\langle \beta_1^1, \beta_2^1, \beta_1^2, \beta_3^1, \beta_2^2, \beta_1^3, \ldots\rangle$ is identical with the sequence $\langle x_1, x_2, x_3, \ldots\rangle$.

The sequence $\{C_n: n = 1, 2, 3, \ldots\}$ of the formulas is defined by the conditions:

$$C_1 = A_1(\alpha_1^1, \beta_1^1),$$

$$C_n = C_{n-1} \& A_1(\alpha_n^1, \beta_n^1) \& A_2(\alpha_{n-1}^2, \beta_{n-1}^2) \& \ldots \& A_{n-1}(\alpha_2^{n-1}, \beta_2^{n-1}) \& A_n(\alpha_1^n, \beta_1^n) .$$

Let us observe that $\{(\alpha_1)(E\beta_1)A_1, \ldots, (\alpha_n)(E\beta_n)A_n\} \vdash (P_n)C_n$, where (P_n) is a string of existential quantifiers binding all the free variables in C_n. If, therefore, every finite set $S_f \subset S$ is satisfiable, so is every C_n. But, if every C_n is satisfiable, so by the same argument used in the proof of lemma 9 we conclude that there is a model of the set S.

3. The first order predicate calculus with identity

We shall obtain the first order predicate calculus with identity after enriching the former language with two-argument predicate symbol $=$ (symbol of identity) and adding to it two axioms:

(A7) $x = x$

(A8) $(x = y) \to (F(x) \to F(y))$, where $F(y)$ is the result of replacing some, but not necessarily all of the free occurrences of x in $F(x)$ by y.

We have used the sign $=$ earlier in the metasystem but it will not be the cause of ambiguity.

The following generalization of theorem I holds.

THEOREM 7. *Every formula of an extended language has the property* (Γ).

Proof. Let A be any formula of the extended language. Formula B is constructed as a conjunction of the formulas: A, $(x)(x = x)$ and of all the formulas that we obtain from A8 by substituting for F the predicate variables occurring in A and by prefixing with a string of universal quantifiers binding all free variables to these formulas. To be more precise, if G is one-argument predicate variable, then as a result of the described operation, we shall get the formula $(x)(y)(x = y) \to [G(x) \to G(y)]$; for two-argument predicate variable H we shall get two formulas:

$$(x)(y)(z)\{(x = y) \to [H(x, z) \to H(y, z)]\}$$

$$(x)(y)(z)\{(x = y) \to [H(z, x) \to H(z, y)]\} .$$

In an analogical way the other formulas are formed by means of predicate variables of more arguments. Let us denote by B' the formula resulting from B by replacing in it every formula of the form $x = y$ by the formula $G_1(x, y)$, where G_1 is two-argument predicate variable not otherwise occurring in B. The formula B' is expressed in the language without identity and, according to what was proved above, it is satisfiable or it is counter-thesis. If B' is a counter-thesis then B also is, because B is a conjunction of the formula A and the other formulas which are theses. By virtue of De Morgan's Law we conclude that $\vdash \bar{A}$. Let us assume that B' is satisfiable in a denumerable model M. From the construction of B' it follows that the relation g_1, corresponding in this model to the symbol G_1, is an equivalence. The quotient model M/g_1 verifies the formula B. The proof of theorem 7 is completed.

Theorem 7 is generalized in an analogical way as theorem 1 to the compactness theorem for the predicate calculus with identity.

At the end his dissertation Gödel examines the independence of these axioms by means of metamathematical argument originating from Bernays [1]. Bernays proved in [1] the independence of (A1)–(A4). The fact, that they remain independent in spite of attaching the axioms (A5)–(A8), is proved by means of an analogical interpretations. Gödel observes that the rules (R1)–(R4) are also independent.

4. Final remarks

The formulas B and C introduced in the course of the proofs of lemma 10 and theorem 4, are patterned on the analogical formulas used by Skolem [7] in proving Löwenheim's theorem. The originality of the Gödel's proof consists in introducing the formulas $(P_n)A_n$. By means of these formulas lemma 8 allows us, in one case, to reduce the completeness problem for the predicate calculus to the completeness theorem for the propositional calculus. A construction of the sequence of the models $\{M^{(n)}\}$ can be also based upon König's infinity lemma [4] which states that every finitely generated tree with infinitely many points must contain at least one infinite branch. König's lemma was becoming to be known among the mathematicians at the time Gödel was writing his dissertation.

References

[1] Bernays P.: *Axiomatische Untersuchungen des Aussagenkalküls der „Principia Mathematica"*.
Mathematische Zeitschrift, 25(1926), 305–320.

[2] Gödel K.: *Die Vollständigkeit der Axiome des logischen Funktionenkalküls.*
Monatshefte für Mathematik und Physik, 37(1930), 349–360.

[3] Hilbert D., Ackermann W.: *Grundzüge der theoretischen Logik.*
Berlin 1928.

[4] König D.: *Sur les correspondances multivoques des ensambles.* Fundamenta Mathematicae, 8(1926), 114–134.
[5] Löwenheim L.: *Über Möglichkeiten im Relativkalkül.* Mathematische Annalen, 76(1915), 447–470.
[6] Post E.: *Introduction to a general theory of elementary propositions.* American Journal of Mathematics, 43(1921), 163–185.
[7] Skolem T.: *Logisch-kombinatorische Untersuchungen über die Erfüllbarkeit oder Beweisbarkeit mathematischer Sätze nebst einem Theoreme über dichte Mengen.* Videnskapsselskapets Skrifter, I. Matematisk-naturvidenskabelig klasse, 4 (1920).

JAN ZYGMUNT

A SURVEY OF THE METHODS OF PROOF OF THE GÖDEL-MALCEV'S COMPLETENESS THEOREM [1]

Introduction

1. The aim of this paper is to give a review of the methods of proving the extended completeness theorem for the lower predicate calculus with identity, which is also called the Gödel-Malcev's theorem. According to this theorem, every consistent set of formulas in the first order language has a model. Two other basic theorems follow the Gödel-Malcev's theorem. Namely, the compactness theorem which states that every set has a model if each of its finite subsets has a model, and also Gödel's narrower completeness theorem according to which every formula in the first order language which is universally valid, is a thesis of the lower predicate calculus with identity.

In the subsequent part of these introductory remarks by formulas we shall mean only the formulas in the first order language.

The theorems quoted above are considered to be the basic theorems of model theory. In model theory investigations are made into the relationship between the mathematical relational-functional structures which are called models and the sets of formulas serving as the axiom sets with respect to the former. More precise formulation of any problem which gives account of the mentioned relationships, requires application of the concept fundamental for theory of models, i.e., the concept of satisfaction which has been introduced by Professor Tarski. These relationships are being established, i.a., by giving various methods of constructing models that verify the given sets of formulas. There are two sound reasons to

---

[1] I am greatly indebted to Professor Stanisław J. Surma who was always ready to give valuable advise and criticism while this paper was being written.

consider the knowledge of (the methods of) construing a model to be valuable. The first reason is related to applications of the Gödel-Malcev's theorem in mathematics, in particular, in algebra where — from model theory emerged. The other reason is motivated by metatheoretic issues and particularly by meta — set-theoretic and meta-arithmetic ones. In this case the ability to construct models allows solving the problems pertaining to consistence and independence of axioms.

The theorems discussed in the present paper were stated, as it often happens with the basic theorems in mathematical disciplines, as early as in the beginning of model theory (1930, 1936). With the course of years there were different versions of their proofs presented. It appears that the majority of these methods is based on the completeness theorem for the classical propositional calculus (1921) or, otherwise, on Lindenbaum's theorem on consistent and complete supersystems (1930). Lindenbaum's theorem appears in subsequent methods in various algebraic versions. All these methods, however, are united by the construction of the reduced ultraproduct of the family of models.

In spite of the fact that the concept of ultraproduct was formulated in an explicit way not earlier than in 1955, it was nevertheless present in an implicit form in the original method of Gödel and likewise in Skolem's construction of the non-standard model for arithmetic (1934), cf. [45]. On the other hand Gödel's constructions reach to the sources of model theory which are to be found in the ideas of Löwenheim (1915) and Skolem (1920), cf. [44].

The review of various methods of proving the Gödel-Malcev's theorem enables us to observe to what degree it inspired the formal development of the lower predicate calculus, and to see its role in contributing to the development of mathematical disciplines related to logic, particularly, the theory of Boolean algebras and their generalizations, i.e., cylindric algebras of Henkin-Tarski and polyadic algebras of Halmos (cf. [14], [15] and [11]).

In 1939 D. Hilbert and P. Bernays ([18], p. 205–253) gave the so called „proof-theoretic" version of the narrower completeness theorem of Gödel. This version of the theorem states as follows: it a formula is irrefutable in the lower predicate calculus, it is irrefutable also in every consistent system T that remains consistent when the axioms of number theory, as well as any other verifiable formulas of the theory, are added to the axioms of T (cf. also [21], p. 395, theorem 36).

The „proof-theoretic" version has been strenghtened by H. Wang cf., e.g., [52], p. 345–352 for the case of infinite set of formulas. H. Wang gave still another version of the discussed theorem of Hilbert-Bernays according to which in arithmetic with the axiom $con(T)$ added, it is possible to prove arithmetical translations of all theorems of the system T. The formula $con(T)$ expresses the consistency of the elementary system T.

2. The present article is essentially a systematic and historic study. We have been trying to give a faithful account of the ideas of particular authors and to give a stress to the originality of their constructions. However, the sequence of arguments as well as the large portion of proofs have undergone deep changes. The proofs of the majority of lemmas are carried on here in a detailed way. Without proof are left some apparent logical lemmas which are quoted for the sake of clarity and also some lemmas generally acknowledged to be fundamental in mathematical logic; finally, some proofs of auxiliary lemmas of purely mathematical character are omitted here.

3. In the present paper we have adopted the formalism of Hilbert's type for the lower predicate calculus cf. [17]; thus the present review is made under this general assumption. This was also the reason to omit completely the problems of completeness of the predicate calculus given by the formalism of Gentzen's type. Neither were the problems of completeness discussed with respect to the predicate calculus conceived of as a system of natural deduction either in the version of Beth's diagrams, or in the version of analytic tableaux of Smullyan. Some questions pertaining to these problems will be noted in the conclusion of the paper.

4. In the present article the use is made of the common concepts and symbols of set theory, topology and algebra. The symbols $=$, $\neq$, $\subset$ denote, respectively, equality, non-equality and inclusion. The symbol $FinZ$ denotes the family of all finite subsets of the set Z. By Z^T we symbolize the set of all functions f such that $f\colon T \to Z$. The set of natural numbers is denoted by N, and the set of all finite ordinal numbers by ω. The symbol „iff“ is an abbreviation of the metasystem expression „if and only if“ while the symbol „non $W(x)$“ abbreviates the expression „it is not true that the condition $W(x)$ is satisfied“.

I. Preliminaries

1. The description of the language of the lower predicate calculus

The alphabet of the lower predicate calculus (written LPC for brevity) is the set that contains as its elements the following sets of symbols:

(i) the infinite set $\{v_n\colon n \in \omega\}$ of individual variables,

(ii) the set $\{c_i\colon i \in I_1\}$, where $I_1 \subset \omega$, of individual constants,

(iii) the set $\{P_i\colon i \in I_2\}$, where $\emptyset \neq I_2 \subset \omega$, of predicate constants,

(iv) the set $\{F_i\colon i \in I_3\}$, where $I_3 \subset \omega$, of function constants,

(v) the set $\{\doteq\}$ which contains the identity symbol $\doteq$ as its only element,

(vi) the set $\{\neg, \wedge, \vee, \rightarrow, \leftrightarrow\}$ of logical connectives, i.e., respectively, negation, conjunction, disjunction, implication, and equivalence,

(vii) the set $\{(v_i), (Ev_i)\}$ which contains as its elements the universal quantifier and the existential quantifier binding the variable v_i,

(viii) the set of punctuation symbols which contains the commas and parentheses.

We establish an arbitrary $\delta \in \omega^{I_2}$ and $\gamma \in N^{I_3}$. We assume that to the predicate constant P_i there corresponds a number $\delta(i)$ called its arness and that to each function constant F_i there corresponds the number $\gamma(i)$ called its arness.

By T we symbolize the set of all terms. We define it as the smallest set satisfying the conditions:

(i) $\{v_i: i \in \omega\} \subset T$,

(ii) $\{c_i: i \in I_1\} \subset T$,

(iii) if $t_1, \ldots, t_{\gamma(i)} \in T$, then $F_i(t_1, \ldots, t_{\gamma(i)}) \in T$.

By At we shall symbolize the set of all atomic formulas. We define it as the smallest set satisfying the conditions:

(i) $t_1 \doteq t_2 \in At$,

(ii) if $t_1, \ldots, t_{\delta(i)} \in T$ then $P_i(t_1, \ldots, t_{\delta(i)}) \in At$.

By S we denote the set of all formulas. We define it as the smallest set satisfying the conditions:

(i) $At \subset S$,

(ii) if $A, B \in S$, then $(\neg A) \in S$, $(A \wedge B) \in S$, $(A \vee B) \in S$, $(A \rightarrow B) \in S$, $(A \leftrightarrow B) \in S$, $(v_i)A \in S$, $(Ev_i)A \in S$.

By the language of the *LPC* we shall understand the triple whose the elements are: alphabet, the set T and the set S.

In further considerations we shall make use of the metasystem notations $x, y, \ldots$ with or without indices, to stand for individual variables. To make writing of the formulas simpler we shall also adopt the familiar conventions of omitting the parentheses.

We also accept the familiar definition of the bound variable and of the free variable.

We shall say that the term t is free for the variable x in the formula A, provided that no free occurrence of x in A appears in the scope of the quantifier binding the variable y of which the term t is composed.

By S_0, where $S_0 \subset S$, we shall denote the set of all open formulas. In other words, S_0 is the set of all formulas without quantifiers.

If a term t is free for a variable x in A, then by $A(x/t)$, or shortly $A(t)$, we shall denote the result of substituting the term t for the variable x in any free occurrence of x in A.

If the formula A contains the variable x free, then we often indicate this situation by $A(x)$.

By S_1, where $S_1 \subset S$, we shall symbolize the set of all sentences, i.e., formulas in which there appears no free variable.

2. The concept of a model for the language of *LPC*

Consider the following relational-functional structure:

$$\mathfrak{M} = \langle U, \{s_i\colon i \in I_1\}, \{R_i\colon i \in I_2\}, \{f_i\colon i \in I_3\}, Id\rangle$$

where I_1, I_2 and I_3 are the sets appearing in the definition of alphabet of *LPC*, in which

(i) $U \neq \emptyset$,

(ii) $\{s_i\colon i \in I_1\} \subset U$,

(iii) $R_i \subset U^{\lambda(i)}$, for some $\lambda \in \omega^{I_2}$,

(iv) $f_i\colon U^{\chi(i)} \to U$, for some $\chi \in N^{I_3}$,

(v) $Id \subset U^2$ and Id is a relation which is reflexive, symmetric and transitive.

We shall say that the relational-functional structure $\mathfrak{M}$ is a model for the language of *LPC*, or shortly a model, provided that there exists a one-to-one function $\mathfrak{z}$ defined on the set $\{c_i\colon i \in I_1\} \cup \{P_i\colon i \in I_2\} \cup \{F_i\colon i \in I_3\} \cup \{\doteq\}$, and such that

(i) $\mathfrak{z}(c_i) = s_i$,

(ii) $\mathfrak{z}(P_i) = R_i$,

(iii) $\mathfrak{z}(F_i) = f_i$,

(iv) $\mathfrak{z}(\doteq) = Id$

and $\lambda(i) = \delta(i)$, for $i \in I_2$ and $\chi(i) = \gamma(i)$, for $i \in I_3$.

We shall say that a model of the language of *LPC* has been defined on a non-empty domain U provided that the set $\{s_i\colon i \in I_1\} \subset U$ is distinguished and that in U there is defined a family $\{R_i\colon i \in I_2\}$ of relations, the family $\{f_i\colon i \in I_3\}$ of functions and equivalence relation Id, such that

$$\langle U, \{s_i\colon i \in I_1\}, \{R_i\colon i \in I_2\}, \{f_i\colon i \in I_3\}, Id\rangle$$

is the model in sense defined above.

We establish that $\mathfrak{M}$ is a model of the language of *LPC* and that U is its domain.

Any function $\bar{u}$ such that

$$\bar{u}: \{v_i: i \in \omega\} \to U$$

will be called a valuation of individual variables in the model $\mathfrak{M}$. Observe that $\bar{u}$ may be identified with the sequence

$$\langle u_0, u_1, \ldots, u_i, \ldots \rangle .$$

Let $\bar{u} \in U^\omega$ and $e \in U$. The symbol $\overline{u(i/e)}$ will denote the valuation given by the following:

$$\overline{u(i/e)} = \langle u_0, \ldots, u_{i-1}, e, u_{i+1}, \ldots \rangle .$$

By $t[\bar{u}]$ we shall symbolize the value of the term t under the valuation $\bar{u}$. We assume inductively that

(i) $v_i[\bar{u}] = u_i$,

(ii) $c_i[\bar{u}] = s_i$, where s_i is an interpretation of the constant c_i and it does not depend on $\bar{u}$,

(iii) $F_i(t_1, \ldots, t_{\gamma(i)})[\bar{u}] = f_i(t_1[\bar{u}], \ldots, t_{\gamma(i)}[\bar{u}])$.

We accept the following definition of the expression „the valuation $\bar{u}$ satisfies the formula A in the model $\mathfrak{M}$“, in symbols

$\mathfrak{M} \models A[\bar{u}]$:

(i) $\mathfrak{M} \models t_1 \doteq t_2[\bar{u}]$ iff $\langle t_1[\bar{u}], t_2[\bar{u}]\rangle \in Id$,

(ii) $\mathfrak{M} \models P_i(t_1, \ldots, t_{\delta(i)})[\bar{u}]$ iff $\langle t_1[\bar{u}], \ldots, t_{\delta(i)}[\bar{u}]\rangle \in R_i$,

(iii) $\mathfrak{M} \models \neg A[\bar{u}]$ iff non $\mathfrak{M} \models A[\bar{u}]$,

(iv) $\mathfrak{M} \models A \wedge B[\bar{u}]$ iff $\mathfrak{M} \models A[\bar{u}]$ and $\mathfrak{M} \models B[\bar{u}]$,

(v) $\mathfrak{M} \models A \vee B[\bar{u}]$ iff $\mathfrak{M} \models A[\bar{u}]$ or $\mathfrak{M} \models B[\bar{u}]$,

(vi) $\mathfrak{M} \models A \leftrightarrow B[\bar{u}]$ iff $\mathfrak{M} \models A[\bar{u}]$ if and only if $\mathfrak{M} \models B[\bar{u}]$,

(vii) $\mathfrak{M} \models A \to B[\bar{u}]$ iff non $\mathfrak{M} \models A[\bar{u}]$ or $\mathfrak{M} \models B[\bar{u}]$,

(viii) $\mathfrak{M} \models (v_i)A[\bar{u}]$ iff for any $e \in U$ $\mathfrak{M} \models A[\overline{u(i/e)}]$,

(ix) $\mathfrak{M} \models (Ev_i)A[\bar{u}]$ iff there exists $e \in U$ such that $\mathfrak{M} \models A[\overline{u(i/e)}]$.

We say of formula A that it is satisfiable in the model $\mathfrak{M}$ provided that there exists a valuation $\bar{u}$ in the model $\mathfrak{M}$ which satisfies it, i.e., there holds the following: $\mathfrak{M} \models A[\bar{u}]$.

We shall say of formula A that it is satisfiable provided that there exists a model $\mathfrak{M}$ in which it is satisfiable.

We shall say of formula A that it is valid in the model provided that every valuation in this model satisfies A.

We shall say of formula A that it is universally valid provided that it is valid in every model.

We shall say that the model $\mathfrak{M}$ is a model of the set of formulas X provided that every formula $A \in X$ is satisfiable in $\mathfrak{M}$.

Let us note now the familiar properties of the concept of satisfaction.

LEMMA 1. *If* $A = A(x_1, \ldots, x_n)$ *is a formula where* $x_1, \ldots, x_n$ *are all free variables and* $\bar{u}_1$, $\bar{u}_2$ *are valuations in the model* $\mathfrak{M}$, *such that* $\bar{u}_1(x_i) = \bar{u}_2(x_i)$, *for* $i = 1, \ldots, n$, *then* $\mathfrak{M} \models A[\bar{u}_1]$ *iff* $\mathfrak{M} \models A[\bar{u}_2]$.

Lemma 1 informs that the satisfaction of a formula by a sequence depends exclusively on the finite number of terms of the sequence. Regarding this, instead of writing $\mathfrak{M} \models A[\bar{u}]$, we shall sometimes write $\mathfrak{M} \models$ $\models A[u_0, \ldots, u_n]$, assuming that all free variables of the formula A are contained in the set $\{v_0, \ldots, v_n\}$. Generally, by writing

$$\mathfrak{M} \models A(x_1, \ldots, x_n)[u_1, \ldots, u_n]$$

we shall understand that the sequence $\bar{u}$ such that $\bar{u}(x_i) = u_i$, for $i = 1, \ldots, n$, satisfies the formula $A(x_1, \ldots, x_n)$ in $\mathfrak{M}$.

LEMMA 2. *If* $A \in S_1$, *then for any model* $\mathfrak{M}$, *from the fact that there exists a valuation* $\bar{u}$ *in the model* $\mathfrak{M}$, *such that* $\mathfrak{M} \models A[\bar{u}]$, *it follows that for any valuation* $\bar{u}$ *in the model* $\mathfrak{M}$ *there holds the following:* $\mathfrak{M} \models A[\bar{u}]$.

Lemma 2 informs about the fact that the concepts of validity and satisfaction of sentences are equivalent.

LEMMA 3. *For any formula* $A(x_1, \ldots, x_n)$ *where* $x_1, \ldots, x_n$ *are all its free variables, and for any model* $\mathfrak{M}$, *the following*

$$\mathfrak{M} \models (Ex_1)\ldots(Ex_n)A(x_1, \ldots, x_n)$$

is equivalent to the existence of such a valuation $\bar{u}$ *in the model* $\mathfrak{M}$, *that* $\mathfrak{M} \models A(x_1, \ldots, x_n)[\bar{u}]$.

The proofs of lemmas 1–3 may be found, e.g., in [10] and [28].

3. Axioms for *LPC*

Let $A, B, D \in S$. Assume the following axioms (A1)–(A17).

(A1) $(A \to B) \to ((B \to D) \to (A \to D))$

(A2) $(A \to (A \to B)) \to (A \to B)$

(A3) $A \to (B \to A)$

(A4) $A \wedge B \to A$

(A5) $A \wedge B \to B$

(A6) $(A \to B) \to ((A \to D) \to (A \to B \wedge D))$

(A7) $A \to A \vee B$

(A8) $B \to A \vee B$

(A9) $(A \to D) \to ((B \to D) \to (A \vee B \to D))$

(A10) $(A \leftrightarrow B) \to (A \to B)$

(A11) $(A \leftrightarrow B) \to (B \to A)$

(A12) $(A \to B) \to ((B \to A) \to (A \leftrightarrow B))$

(A13) $(\neg A \to \neg B) \to (B \to A)$

(A14) $(x)A(x) \to A(t)$, with the assumption that the term t is free for x in $A(x)$,

(A15) $(x)(A \to B) \to (A \to (x)B)$, with the assumption that x is not free in A,

(A16) $(x)(x \doteq x)$

(A17) $(x)(y)((x \doteq y) \to (A \to A'))$, where A' results from A by substituting in some places (not necessarily all), the variable y for the free variable x provided that A is a formula in which y is free for x.

The rules of inference are the following:

(R1) The rule of detachment: B follows from A and $A \to B$,

(R2) The rule of generalization: $(x)A$ follows from A.

Axioms (A1)–(A15) together with (R1) and (R2) characterize the *LPC* without identity. Axioms (A16) and (A17) are the specific axioms for identity.

The list of axioms given above is composed of the axioms for the classical propositional calculus (A1)–(A13) given in [32], p. 32 and the specific axioms for quantification (A14) and (A15) given in [28], p. 57.

The fact that formula A has a proof on the grounds of the set X of formulas will be noted thus: $X \vdash A$. In the case when $X = \emptyset$ the inscription $X \vdash A$ is reduced to $\vdash A$, and it stands for the fact that A is a thesis of *LPC*. If $\vdash \neg A$, then A will be called a refutable formula in *LPC*.

By *Syst* we shall denote the family of all deductive systems contained in the set of all formulas, i.e., the family of all sets X such that $X = \{A \in S: X \vdash A\}$.

By *Con* we shall symbolize the family of all consistent sets included in S, i.e., the family of all sets X such that $\{A \in S: X \vdash A\} \neq S$.

By *Com* we shall understand the family of all complete sets included in S, i.e., the family of all sets X such that for any $A \in S$, $X \vdash A$ or $X \vdash \neg A$.

We shall often refer to the famous Lindenbaum's theorem which is quoted below (cf. Tarski [50]):

LEMMA 4. *If* $X \in Con$, *then there exists a set* X^+ *such that* $X \subset X^+$ *and* $X^+ \in Syst \cap Con \cap Com$.

We shall say of formula $(Q_1x_1)(Q_2x_2)...(Q_nx_n)A$ that it is in a normal form provided that $A \in S_0$, (Q_ix_i) is the universal or existential quantifier binding the variable x_i and $x_i \neq x_j$ for $i \neq j$.

Let us note the lemma which is important in view of further applications:

LEMMA 5. *If* $A \in S$, *then there exists* $B \in S$ *in normal form and such that* $\vdash A \leftrightarrow B$.

The model in which the relation Id is the identity is called the model with an absolute concept of identity. The so called normalization procedure which amounts to the construction of a quotient model allows for obtaining the model with an absolute concept of identity from any model. Consider the model

$$\mathfrak{M} = \langle U, \{s_i\colon i \in I_1\}, \{R_i\colon i \in I_2\}, \{f_i\colon i \in I_3\}, Id\rangle .$$

In the domain $U' = U/Id$, where $u^{\sim} \in U'$ and $u^{\sim} = \{s \in U\colon \langle u, s\rangle \in Id\}$ we define the model

$$\mathfrak{M}' = \langle U', \{s'_i\colon i \in I_1\}, \{R'_i\colon i \in I_2\}, \{f'_i\colon i \in I_3\}, Id'\rangle$$

where

(i) $s'_i = s_i^{\sim}$

(ii) $\langle u_1^{\sim}, ..., u_{\delta(i)}^{\sim}\rangle \in R'_i$ iff $\langle u_1, ..., u_{\delta(i)}\rangle \in R_i$

(iii) $f'_i(u_1^{\sim}, ..., u_{\gamma(i)}^{\sim}) = f_i(u_1, ..., u_{\gamma(i)})^{\sim}$

(iv) $\langle u_1^{\sim}, u_2^{\sim}\rangle \in Id'$ iff $\langle u_1, u_2\rangle \in Id$.

In view of the thesis

$$\vdash x_1 \doteq y_1 \wedge ... \wedge x_{\delta(i)} \doteq y_{\delta(i)} \to (P_i(x_1, ..., x_{\delta(i)}) \leftrightarrow P_i(y_1, ..., y_{\delta(i)}))$$

and in view of the thesis

$$\vdash x_1 \doteq y_1 \wedge ... \wedge x_{\gamma(i)} \doteq y_{\gamma(i)} \to (F_i(x_1, ..., x_{\gamma(i)}) \doteq F_i(y_1, ..., y_{\gamma(i)}))$$

it follows that the relations R'_i and the functions f'_i are correctly defined in the model $\mathfrak{M}'$ which is thus the model with an absolute concept of identity.

If $\bar{u} = \langle u_0, u_1, ..., u_i, ...\rangle$, then by $\overline{u^{\sim}}$ we shall symbolize the sequence $\langle u_0^{\sim}, u_1^{\sim}, ..., u_i^{\sim}, ...\rangle$.

With these notations accepted, there holds the following lemma which states that the content of any model equals to the content of its quotient with respect to the relation Id.

LEMMA 6. *If* $\bar{u}$ *is a valuation in the model* $\mathfrak{M}$, *then for any* $A \in S$ *there holds the equivalence:*

$$\mathfrak{M} \models A[\bar{u}] \quad \textit{iff} \quad \mathfrak{M}' \models A[\overline{u^{\sim}}] .$$

4. Formulation of the completeness theorems

We shall be interested in the methods of proving the completeness theorems stated below.

THEOREM 1 (the narrower completeness theorem, Gödel [9]). *Each universally valid formula is a thesis of LPC.*

THEOREM 2 (Gödel [9]). *Each formula is either refutable in LPC, or it is satisfiable in a denumerable domain.*

THEOREM 3 (the extended completeness theorem, Gödel [9], Malcev [26]) *Every consistent set of formulas has a model.*

THEOREM 4. *Every consistent set of sentences has a model.*

THEOREM 5 (the general principle of localization, or the compactness theorem, Gödel [9], Malcev [27]). *If every finite subset of a certain set of sentences has a model, then the whole set has a model.*

THEOREM 6 (the compactness theorem for arithmetical classes, Tarski [51], Rasiowa [33]). *If* $X = \bigcup_{i\in I} X_i$ *is a set of sentences, and for every* $I_f \in Fin\, I$ *there exists a model of the set* $\bigcup_{i\in I_f} X_i$, *then there exists a model of the set* X.

The subsequent lemma establishes the connections among the quoted completeness theorems.

LEMMA 7.

(a) (1) *is equivalent to* (2)

(b) (3) *is equivalent to* (4)

(c) (5) *is equivalent to* (6)

(d) (3) *is equivalent to the conjunction of* (1) *and* (5).

Proof. The easy proof is omitted.

There also holds an easily provable theorem that is the reverse of theorem 1, and which is called the compatibility theorem.

Until the end of this article we shall restrict ourselves to considering the basic language which will be called the language L and which is obtained from the formerly described one by excluding certain items from its alphabet. The alphabet of L contains individual variables, two predicate constants P and Q, being a unary and a binary one, respectively, and also the symbol of identity. This limitation is motivated merely by didactic and practical reasons and it by no means diminishes the generality of considerations. When we pass over to discussing particular methods, there will be instances of the need appearing to enrich the language L by adding to it function constants and individual constants. This will effect the necessity of following quite general lines in reasoning. The models of our language will be symbolized by $\langle U, P, Q\rangle$ or by $\langle U, P, Q, Id\rangle$.

5. The proof of the narrower completeness theorem for open formulas without identity

The considerations are auxiliary in their character and the use will be made of them in Chapter II. The narrower completeness theorem for open formulas amounts to a certain modification of the completeness theorem for the classical propositional calculus, which states that every tautology in two element Boolean algebra is a thesis of the classical propositional calculus.

By $\mathfrak{S}_0 = \langle S_0, \neg, \wedge, \vee, \rightarrow, \leftrightarrow\rangle$ we shall symbolize a Lindenbaum's algebra of open formulas S_0, where the set of free generators is the set of all atomic formulas At.

Let $Var = \{p_i\colon i \in \omega\}$ be a finite set of propositional variables. By $\bar{S}_0$ we shall symbolize the set of all formulas of the propositional calculus which are composed of the variables p_i with the help of the connectives $\neg, \wedge, \vee, \rightarrow, \leftrightarrow$. Like previously, by $\bar{\mathfrak{S}}_0 = \langle \bar{S}_0, \neg, \wedge, \vee, \rightarrow, \leftrightarrow\rangle$ we shall denote Lindenbaum's algebra of formulas $\bar{S}_0$. The set Var is the set of free generators of this algebra.

From the definition of the set At it follows that there exists a bijection g such $g\colon At \rightarrow Var$. We shall give two obvious lemmas.

LEMMA 8. *A bijection $g\colon At \rightarrow Var$ may be extended to an isomorphism $\bar{g}$ of the algebra $\mathfrak{S}_0$ onto the algebra $\bar{\mathfrak{S}}_0$.*

LEMMA 9. *Every function $v\colon Var \rightarrow \{0, 1\}$ may be extended to the homomorphism $\bar{v}$ of the algebra $\bar{\mathfrak{S}}_0$ onto two-element Boolean algebra.*

The extension $\bar{v}$ mentioned in lemma 9 will be called the Boolean valuation of the set $\bar{S}_0$.

LEMMA 10. *If*

(1) $g\colon At \rightarrow Var$ *is a bijection,*

(2) $\mathfrak{M} = \langle U, P, Q\rangle$ *is a model of the language L without identity,*

(3) $\bar{u} \in U^{\omega}$ *is a valuation of individual variables,*

(4) *the function* $v\colon Var \to \{0, 1\}$ *is defined as follows:*

$$vgP(v_i) = 1 \quad iff \quad \mathfrak{M} \models P(v_i)[\overline{u}]$$

$$vgQ(v_i, v_j) = 1 \quad iff \quad \mathfrak{M} \models Q(v_i, v_j)[\overline{u}]$$

(5) $A \in S_0$,

then

(6) $\mathfrak{M} \models A[\overline{u}]$ *iff* $\overline{v}\overline{g}A = 1$.

Proof. We carry on the induction on the construction of the A. Assume (1)–(5).

(I) If $A \in At$, then the (6) is satisfied in view of (4).

(II) Assume that

(7) the condition (6) is satisfied for the formulas B and D.

Let it be

(8) $\mathfrak{M} \models \neg B[\overline{u}]$.

From the definition of satisfaction it follows that (8) is equivalent to the fact that

(9) non $\mathfrak{M} \models B[\overline{u}]$.

From the inductive assumption (7) it follows that (9) is equivalent to the fact that

(10) $\overline{v}\overline{g}B = 0$.

From the fact that v and g are homomorphisms, we infer that (10) is equivalent to the fact that

(11) $\overline{v}\overline{g}\neg B = 1$.

Hence and from (8) it follows that the equivalence (6) holds for the formulas of the forms $\neg B$. Along the same lines we prove that the condition (6) holds for the formulas of the form $B \wedge D$ what is sufficient condition for thesis of lemma 10.

Q. E. D.

Observe that the assumption (1) is utilized in the proof, viz., in the point (4) we use it for correct defining the function v.

LEMMA 11. *If*

(1) $g\colon At \to Var$ *is a bijection,*

(2) $A \in S_0$,

(3) *there exists* $v\colon Var \to \{0, 1\}$ *such that* $\overline{v}\overline{g}A = 1$,

then there exists a model $\mathfrak{N}$ *in which* A *is satisfied.*

Proof. Assume (1)–(3). The model $\mathfrak{N} = \langle U, P, Q\rangle$ is defined as follows:

(i) $U = \omega$,

(ii) $i \in P$ iff $vgP(v_i) = 1$,

(iii) $\langle i, j\rangle \in Q$ iff $vgQ(v_i, v_j) = 1$.

Analogously as in lemma 10, we prove that there hold the following:

(4) $\mathfrak{M} \models A[0, 1, 2, \ldots]$ iff $\overline{vg}A = 1$

wherefrom the thesis of the lemma follows.

Q. E. D.

By the symbol $\underset{0}{\vdash} A$ we shall denote the fact that in the proof of A, we have based only on the axioms (A1)–(A13) and the rule (R1).

LEMMA 12. *If* $A \in S_0$, *then* $\vdash A$ *if and only if* $\underset{0}{\vdash} A$.

Proof. Assume that $A \in S_0$ and that

(1) $\vdash A$.

Assume indirectly that

(2) non $\underset{0}{\vdash} A$.

Assume moreover, that

(3) $g\colon At \to Var$ is a bijection.

Observe that it follows from (2) and (3), that

(4) $\bar{g}A$ does not result from any tautology of the classical propositional calculus.

It follows from (4) that there exists $v\colon Var \to \{0, 1\}$ such that

(5) $\overline{vg}A = 0$.

From (5) and lemma 11 it results that

(6) there exists a model $\mathfrak{N}$ in which $\neg A$ is satisfied.

It follows from (1) by the compatibility theorem that

(7) in model $\mathfrak{N}$ formula A is satisfied.

Hence and from (6) it follows that (2) cannot be accepted.

The reverse implication in the thesis of the lemma is obvious.

Q. E. D.

LEMMA 13 (the narrower completeness theorem for open formulas). *If* $A \in S_0$, *then from the fact that* A *is a universally valid formula, it follows that* $\vdash A$.

Proof. Assume that

(1) $A \in S_0$ and A is a universally valid formula.

Suppose indirectly that

(2) non $\vdash A$.

From (2) and lemma 12 it follows that

(3) non $\vdash_0 A$.

Hence we infer that for any bijection $g \colon At \to Var$

(4) $\bar{g}A$ is not a tautology of the classical propositional calculus.

This yields in turn that there exists the function $v \colon Var \to \{0, 1\}$ and such that

(5) $\bar{v}\bar{g}A = 0$.

This together with lemma 11, in view of the fact that $\bar{v}$ and $\bar{g}$ are homomorphisms, yields that

(6) there exists a model in which $\neg A$ is satisfied.

Condition (6) contradicts line (1), thus we have that

(7) $\vdash A$.

Q. E. D.

The method of proving lemma 13 that was here chosen caused a limitation; we give only the narrower completeness theorem valid for the open formulas without identity. The method, originated from [9] and [30], based on the completeness theorem for the classical propositional calculus cannot be immediately transferred to the *LPC* with identity since in this calculus there are atomic formulas which are theses, while in the classical propositional calculus there are no propositional variables which would be theses.

This fact does not by any means exclude the possibility of giving a proof of lemma 13 for the open formulas in *LPC* with identity. The accepting of the Łoś' method (cf. [23]) leads to solving the problem. We have, however, given up that way out since we aim at a faithful account of the constructions of Gödel, Mostowski and Beth.

II. Mostowski's method of proving the narrower completeness theorem for the *LPC* without identity

The method of Mostowski which was published in [30] in 1948 is based on the completeness theorem for open formulas and on the topological interpretation of the concept of satisfaction in Cantor discontinuum.

This presentation of the method which is given in [30] is overloaded with longish and quite complicated proofs. This chapter aims at giving a more concise account of Mostowski's method with simultaneous modification of the topological apparatus there employed and making references to Gödel's results which will give a stress to the connections between the two methods.

Observe that A. Mostowski discussed the possibility of applying topology to prove completeness theorems for logical calculi in his speech at the Tenth International Congress for Philosophy in Amsterdam in 1948 (cf. [31]).

1. Topological interpretation of the concept of satisfaction

The set $\mathfrak{C} = \{0, 1\}^N$ is called the Cantor set. $\mathfrak{C}$ contains the sequences of the type $\langle s_q : q \in N\rangle$, where $s_q = 1$ or $s_q = 0$.

LEMMA 1. *For every* $n \in N$ *there exists a bijection mapping the set of all n-ary relations defined in* N *onto the set* $\mathfrak{C}$.

Proof. Assume that $n \in N$. It is familiar that the cardinality of the sets N and N^n is the same. Then there exists a bijection

$$h_n : N \to N^n .$$

We define the function $\bar{h}_n : 2^{N^n} \to \mathfrak{C}$ as follows. Let $R \in 2^{N^n}$. Let us put or any $q \in N$

$$s_q = \begin{cases} 1, & \text{if} \quad h_n(q) \in R \\ 0, & \text{if} \quad h_n(q) \notin R . \end{cases}$$

The function $\bar{h}_n$ is defined by the following:

$$\bar{h}_n(R) = \langle s_q : q \in N\rangle .$$

It is easy to check that $\bar{h}_n$ is a one-to-one function „in". We shall prove, moreover, that $\bar{h}_n$ is a surjection. For this purpose let us assume that $\xi = \langle \bar{s}_q : q \in N\rangle \in \mathfrak{C}$. The point ξ determines exactly one n-ary relation R_ξ defined in N by means of the following condition:

$$R_\xi = \{\langle m_1, \ldots, m_n\rangle \in N^n : \bar{s}_{h_n^{-1}(m_1, \ldots, m_n)} = 1\} .$$

It may be easily checked that $\bar{h}_n(R_\xi) = \xi$.

Q. E. D.

When $\bar{h}_n$ is established we shall symbolize the value of $\bar{h}_n(R)$ by ξ_R.

Lemma 1 was given in its full generality. It may be applied in considerations regarding the language with predicate constants of any arness.

Regarding the fact that in our restricted language L there are only two predicate constants P and Q which are respectively a unary and a binary one, we establish for the purpose of further considerations the bijections:

$h_1 \colon N \to N$,

$h_2 \colon N \to N^2$.

When the bijections h_1 and h_2 are selected in this way, each pair $\langle \xi_1, \xi_2 \rangle \in \mathfrak{C}^2$ determines a model

$$\mathfrak{N} = \langle N, P_{\xi_1}, Q_{\xi_2} \rangle$$

of the language L. Conversely, every model $\mathfrak{N} = \langle N, P, Q \rangle$ determines the pair $\langle \xi_P, \xi_Q \rangle \in \mathfrak{C}^2$.

Let $A \in S$, and let $\overline{u} \in N^\omega$. Let us also put that

$$\mathfrak{C}_A[\overline{u}] = \{\langle \xi_1, \xi_2 \rangle \in \mathfrak{C}^2 \colon \langle N, P_{\xi_1}, Q_{\xi_2} \rangle \models A[\overline{u}]\} .$$

For instance

$$\mathfrak{C}_{P(v_i)}[\overline{u}] = \{\langle \xi_1, \xi_2 \rangle \in \mathfrak{C}^2 \colon u_i \in P_{\xi_1}\} .$$

The set $\mathfrak{C}_A[\overline{u}]$ is in a sense a diagram of the formula A in the space $\mathfrak{C}^2$. In view of the above considerations the following is valid:

LEMMA 2.

(a) *The existence of a model in the domain N, in which A is satisfied, is equivalent to the fact that there exists such $\overline{u} \in N^\omega$ that $\mathfrak{C}_A[\overline{u}] \neq \emptyset$.*

(b) *Formula A is valid in every model in the domain N provided that for every $\overline{u} \in N^\omega$ there holds the equality: $\mathfrak{C}_A[\overline{u}] = \mathfrak{C}^2$.*

Let us now recall a few facts from general topology.

Let $n \in N$. Assume that $\mathfrak{C}(n) = \{\xi \in \mathfrak{C} \colon \xi(n) = 1\}$. Let us, moreover, put that $\mathfrak{B}_0 = \{\mathfrak{C}(n) \colon n \in N\}$. By $\mathfrak{B}$ we shall symbolize the smallest field of sets generated by $\mathfrak{B}_0$. We shall consider the set $\mathfrak{C}$ as a topological space assuming that $\mathfrak{B}_0$ is its subbasis. Thus obtained topological space is called Cantor discontinuum.

LEMMA 3. *The pair $\langle \mathfrak{C}, \mathfrak{B} \rangle$ is a compact topological space. The field $\mathfrak{B}$ is the family of all clopen sets included in $\mathfrak{C}$.*

Proof. Cf., e.g., [35], p. 91–92.

Observe furthermore that the power $\mathfrak{C}^2$ with Tychonoff product topology is a compact topological space. This results from the famous Tychonoff theorem (cf., e.g., [6], p. 112).

Let us note still one remark. In each product $T \times U$ with Tychonoff product topology, of the topological spaces T and U, there holds the equality $Int(A \times B) = Int A \times Int B$, where $A \subset T$, $B \subset U$ and Int is the interior operation (cf., e.g., [6], p. 82, exercise D).

Lemma 3 as well as the remarks which follow it will be utilized in a brief proof of lemma 4 and also in the considerations of the subsequent chapter.

LEMMA 4. *If* $A \in S_0$, *then for any* $\overline{u} \in N^\omega$, *the set* $\mathfrak{C}_A[\overline{u}]$ *is clopen in the space* $\mathfrak{C}^2$.

Proof. The proof is carried on by induction on the construction of A.

I. Let $A = P(v_i)$. From the definition of the set $\mathfrak{C}_A[\overline{u}]$ it results that

$$\mathfrak{C}_{P(v_i)}[\overline{u}] = \{\langle \xi_1, \xi_2 \rangle \in \mathfrak{C}^2 \colon \langle N, P_{\xi_1}, Q_{\xi_2} \rangle \models P(v_i)[\overline{u}]\} = \\ = \{\langle \xi_1, \xi_2 \rangle \colon u_i \in P_{\xi_1}\} = \{\xi_1 \colon u_i \in P_{\xi_1}\} \times \mathfrak{C} .$$

Consider the set $Z = \{\xi_1 \colon u_i \in P_{\xi_1}\}$. We shall demonstrate that $Z \in \mathfrak{B}_0$. Actually, the condition $u_i \in P_{\xi_1}$ is equivalent to the fact that $s_{n_0} = 1$, where $n_0 = h_1^{-1}(u_i)$ and s_{n_0} is the n_0-th coordinate of the point ξ_1. Thus $Z = \mathfrak{C}(n_0)$ and hence, $Z \in \mathfrak{B}_0$. From lemma 3 it follows that Z is a clopen set in $\mathfrak{C}$. Hence and from the remarks following lemma 3, we infer that $Z \times \mathfrak{C}$, i.e. $\mathfrak{C}_{P(v_i)}[\overline{u}]$ is a clopen set in $\mathfrak{C}^2$.

An analogous reasoning yields that the set $\mathfrak{C}_{Q(v_i, v_j)}[\overline{u}]$ is a clopen set in $\mathfrak{C}^2$.

II. The inductive step is easy. To make it we ought to base on the following apparent equalities:

$$\mathfrak{C}_{\neg A}[\overline{u}] = \mathfrak{C} - \mathfrak{C}_A[\overline{u}] ,$$
$$\mathfrak{C}_{A \wedge B}[\overline{u}] = \mathfrak{C}_A[\overline{u}] \cap \mathfrak{C}_B[\overline{u}]$$

and also on the fact that both the complement of a clopen set and the product of such sets are the sets of the same kind.

Q. E. D.

2. The proof of the narrower completeness theorem

Throughout this chapter we shall restrict ourselves to considering the formula

$$B_0 = (x)(Ey)A_0(x, y)$$

in the normal form (which is, at the same time, the form of Skolem type). This restriction does not diminish the generality of consideration since the same might be done for any formula in the normal form. Choosing of B_0 will allow us to expose all the essential points of Mostowski's method and to demonstrate the apparent connection of this method with the Gödel's.

Assume also that the sequence of all individual variables has been re-numbered into the sequence $\langle x_1, x_2, \ldots\rangle$.

We still reserve the symbol f to denote the function f such that

$$f: N \to N .$$

LEMMA 5. *If* $\mathfrak{M} = \langle N, P, Q\rangle$ *is a model of the language* L, *then we say that*

$$\mathfrak{M} \models B_0$$

is true provided that there exists such f *that for all* $m \in N$ *there holds the following:*

$$\mathfrak{M} \models A_0(x, y)[m, f(m)] .$$

Proof. The simple proof based on the definition of satisfaction is omitted.

Assume that the function f that we have chosen satisfies the two conditions:

(i) if $m \neq n$, then $f(n) \neq f(m)$,

(ii) $m < f(m)$.

Let $s \in N$. The model $\mathfrak{N}^{(s)} = \langle N, P^{(s)}, Q^{(s)}\rangle$ will be called the model of the degree s for the formula B_0, provided that for any $m \leqslant s$ the following is true:

$$\mathfrak{N}^{(s)} \models A_0(x, y)[m, f(m)] .$$

LEMMA 6. *If for any* $s \in N$ *there exists a model* $\mathfrak{M}^{(s)}$ *of the degree* s *for the formula* B_0, *then there exists a model of the sentence* B_0 *in the domain* N.

Proof. Let us accept the assumptions of the lemma. Let us also put that

$$\mathfrak{C}_s = \bigcap_{m=1}^{s} \mathfrak{C}_{A_0}[m, f(m)] .$$

From the definition of the set $\mathfrak{C}_s$ it follows that for any $s \in N$:

(1) $\mathfrak{C}_s$ is a non-empty set since from the assumption that there exists a model of the degree s for B_0, it results that $\langle \xi_{P^{(s)}}, \xi_{Q^{(s)}}\rangle \in \mathfrak{C}_s$.

(2) $\mathfrak{C}_s$ is closed set in $\mathfrak{C}^2$ being the finite product of the closed sets in $\mathfrak{C}^2$.

(3) $\mathfrak{C}_s \supset \mathfrak{C}_{s+1}$.

From (1)–(3) and from Cantor's theorem which states that: in a compact topological space the set-theoretical product of every descending sequence of non-empty, closed sets, is non-empty, we infer that there exists a pair $\langle \xi_1, \xi_2 \rangle \in \bigcap_{s \in N} \mathfrak{C}_s$.

Since $\langle \xi_1, \xi_2 \rangle \in \mathfrak{C}_{A_0}[m, f(m)]$, for any $m \in N$, then

$$\langle N, P_{\xi_1}, Q_{\xi_2} \rangle \models A_0(x, y)[m, f(m)] .$$

Hence and from lemma 5 we deduce that $\langle N, P_{\xi_1}, Q_{\xi_2} \rangle \models B_0$.

Q. E. D.

By $\bar{A}_s$ we shall symbolize the following disjunction:

$$\bar{A}_s = \neg A_0(x_1, x_{f(1)}) \vee \ldots \vee \neg A_0(x_s, x_{f(s)}) .$$

LEMMA 7. *If there exists such $s \in N$ that there exists no model $\mathfrak{N}^{(s)}$ of the degree s for B_0, then $\vdash \bar{A}_s$.*

Proof. Assume that

(1) there exists no model $\mathfrak{N}^{(s)}$ for B_0

and let us indirectly suppose that

(2) non $\vdash \bar{A}_s$.

Observe that $\bar{A}_s \in S_0$. Hence and from (2) by the completeness theorem for open formulas (lemma 13 and lemma 11, Chapter I), we deduce that there exists a model $\mathfrak{N} = \langle N, P, Q \rangle$ such that

(3) $\mathfrak{M} \models \neg \bar{A}_s[1, 2, \ldots]$.

Since $\neg \bar{A}_s$ is equivalent to a certain conjunction then each conjunct $A_0(x_m, x_{f(m)})$ of this conjunction is satisfiable in the model $\mathfrak{N}$; putting it more exactly, for any $n \leqslant s$ the following holds:

(4) $\mathfrak{N} \models A_0(x_m, x_{f(m)})[m, f(m)]$.

Contrary to (1), it follows from (4) that

(5) $\mathfrak{N}$ is the model of the degree s for B_0 which completes the proof.

Q. E. D.

Let us pass over now to the conclusive fragment of the proof of the narrower completeness theorem. In the case when for every s there exists a model of the degree s for B_0, we infer on the grounds of lemma 6 that B_0 is satisfiable. In the opposite case we deduce, in view of lemma 7, that $\vdash \bar{A}_s$. And hence we conclude that $\vdash (x_1) \ldots (x_s) \bar{A}_s$. From the last thesis on the grounds of lemma 8 in [53] we infer that $\vdash \neg (x)(Ey) A_0(x, y)$, i.e., that B_0 is refutable formula.

III. Beth's method of proving extended completeness theorem for *LPC* without identity

The method of E. W. Beth [3] for proving the completeness theorem is the next version of the Gödel's original method. There is a reference made in it to the completeness theorem for the classical propositional calculus. That theorem is referred to in the following formulation:

If a set of sentences is consistent, then there exists a Boolean valuation which satisfies this set.

E. W. Beth generalizes the theorem to cover the case of the so-called normal valuations which, roughly speaking, determine the values of all the formulas with the formulas with quantifiers included. In order to prove the lemma on the existence of normal valuations E. W. Beth refers to the compactness of the space of all Boolean valuations. The compactness of this space results from the fact that it is homeomorphic with the Cantor discontinuum.

Lemma 3 quoted below, which corresponds to lemma 8 of the Gödel's method discussed in [53], strongly stresses the relationship between the two methods. On the other hand, however, the concept of normal valuation brings E. W. Beth's method closer to the method of L. Henkin.

1. Topological space of Boolean valuations

The set $W = \{0,1\}^{Var}$ may be identified with the Cantor's set $= \{0,1\}^{N}$. Let

$$W(p_i) = \{w \in W\colon\ w(p_i) = 1\}$$

and

$$\mathfrak{B}_0 = \{W(p_i)\colon\ p_i \in Var\}\,.$$

By $\mathfrak{B}$ we shall symbolize the smallest field of sets generated by $\mathfrak{B}_0$. We shall regard W as a topological space assuming $\mathfrak{B}_0$ to be its subbasis.

Similarly as in Chapter I, we shall symbolize by $\mathfrak{S}_0$ the Lindenbaum's algebra of the formulas of the propositional calculus, the algebra being generated by the set *Var*.

Let $\overline{w}$ be a homomorphism of the algebra $\mathfrak{S}_0$ onto the two-element Boolean algebra $\mathfrak{K}_0$, i.e., let $\overline{w}$ be an arbitrary Boolean valuation of the set $\mathfrak{S}_0$. By $\overline{W}$ we shall symbolize the set of all valuations $\overline{w}$. It is familiar that the mapping $g\colon W \to \overline{W}$ defined by the condition $g(w) = \overline{w}$ is one-to-one mapping. Assume that for any $a \in \overline{S}_0$

$$\overline{W}(a) = \{\overline{w} \in \overline{W}\colon\ \overline{w}(a) = 1\}\,.$$

In view of the lemma 3 in Chapter III and the above remark, we infer that $\langle \overline{W}, \{\overline{W}(a)\colon\ a \in \overline{S}_0\}\rangle$ is a compact topological space, in which the sets $\overline{W}(a)$ are the only clopen sets included in $\overline{W}$.

From the completeness theorem for the classical propositional calculus and from the definition of the set $\overline{W}(a)$, there results the following.

LEMMA 1.

(a) $\vdash \neg a$ *iff* $\overline{W}(a) = \emptyset$

(b) $\overline{W}(a \wedge b) = \overline{W}(a) \cap \overline{W}(b)$

(c) $\overline{W}(\neg a) = \overline{W} - \overline{W}(a)$

Let $C = \{c_i : i \in \omega\}$ be the infinite set of individual constants. By $L(C)$ we shall symbolize the language obtained from language L by means of enriching it with the set C.

Assume that a variable x is free in the formula $A(x)$. We shall apply the name of reduction of the formula $A(x)$ (in the set C) with respect to the variable x, to each of the formulas $A(x/c_i)$, where $c_i \in C$; such formulas will be symbolized for brevity, by $A(c_i)$.

Every formula obtained by reduction of the formula A with respect to each of its free variables will be called the reduction of the formula A.

The set of formulas which includes as its elements all the reductions of the formulas belonging to some set will be called the reduction of this set.

By S^+ we shall denote the reduction of the set S of all formulas in the language L without identity. The elements of the set S^+ are, e.g., the formulas $P(c_i)$, $(x)P(x)$, $(Ex)Q(x, x)$, $(y)(x)[P(x) \wedge Q(x, y)]$ etc. Observe that $S_1 \subset S^+$, where S_1 is the set of all sentences in the language L.

LEMMA 2. *The following Lindenbaum algebras* $\langle S^+, \neg, \wedge \rangle$ *and* $\langle \bar{S}_0, \neg, \wedge \rangle$ *are isomorphic.*

Proof. By At^+, where $At^+ \subset S^+$, let us symbolize the set of all these formulas in the language $L(C)$, which contain no free variables and such that any logical connective is the main connective in any of them. For instance $(x)P(x)$, $(x)[Q(x, c_i) \wedge P(c_j)]$ belong to the set At^+, but $(x)Q(x, c_i) \wedge P(c_j)$ do not. It is easily seen that At^+ is the set of all free generators of the algebra $\langle S^+, \neg, \wedge \rangle$. Since there exists a bijection g: $Var \to At^+$, then these algebras are isomorphic being the similar free algebras. An extension $\bar{g}$ of the bijection g defined by the following:

$$\bar{g}(p_i) = g(p_i)$$

$$\bar{g}(\neg a) = \neg \bar{g}(a)$$

$$\bar{g}(a \wedge b) = \bar{g}(a) \wedge \bar{g}(b)$$

is the isomorphism which was looked for.

Q. E. D.

Let $\overline{W}^+$ be the set of all Boolean valuation of the set S^+. In view of the remark before lemma 1 and lemma 2 we deduce that the pair $\langle \overline{W}^+, \{W(A)\colon A \in S^+\}\rangle$, where $W(A) = \{\overline{w} \in \overline{W}^+\colon w(A) = 1\}$, is a compact topological space in which the sets $W(A)$ are the only clopen sets included in $\overline{W}^+$.

2. The lemma on existence of normal valuations. Construction of the model

Valuation $\overline{w} \in \overline{W}^+$ will be called normal provided that for every formula of the form $(x)A(x)$ where $(x)A(x) \in S^+$ there holds the condition:

$$\overline{w}\big((x)A(x)\big) = 1 \quad \text{iff} \quad \overline{w}\big(A(c_i)\big) = 1, \quad \text{for every} \quad c_i \in C.$$

Let us consider the set of all formulas in the language L, in which v_0 occurs as a free variable. The variable v_0 will be symbolized by v. Reduce now this set with respect to each of the variables $v_1, v_2, v_3, \ldots$ Let

(+) $\{A_1(v), A_2(v), A_3(v), \ldots, A_n(v), \ldots\}$

be the set of all formulas obtained by means of this reduction. Reduce then the set (+) with respect to the variable v. Observe that for any $n \in N$:

$$\{A_n(c_i)\colon i \in \omega\} \subset S^+ \text{ and also } (v)A_n(v) \in S^+ \text{ and } (Ev)A_n(v) \in S^+ .$$

Obviously, there exists such a formula $B_1 = B(v, v_{i_1}, \ldots, v_{i_n})$ in which the variables $v, v_{i_1}, \ldots, v_{i_n}$ are free for the variable v_i, and at the same time, the formula is that $A_n(v)$ results from B_1 by reduction of the latter with respect to $v_{i_1}, \ldots, v_{i_n}$.

To each formula $A_n(c_i) \in S^+$ we will assign a formula $A_n^0(v_i) = B(v/v_i, v_{i_1}/v_i, \ldots, v_{i_n}/v_i)$. Observe that $A_n^0(v_i) \in S$.

The function

$$s\colon N \to N$$

is defined by the conditions:

$s(1)$ = the smallest natural number which is larger than any index in the formula $A_1(v)$; $A_1(v)$ belongs to the set (+).

$s(q+1)$ = the smallest natural number which is larger than each index in the formula $A_{q+1}(v)$ and is larger than $s(q)$.

With these notations accepted the following is valid.

LEMMA 3. *For any q and $\overline{q} \in N$*

$$\vdash (Ev_{s(1)})(Ev_{s(2)})\ldots(Ev_{s(\overline{q})}) \bigwedge_{q \leqslant \overline{q}} [A_q^0(v_{s(q)}) \to (v)A_q^0(v)] ,$$

where $\bigwedge_{q\leqslant\overline{q}}$ is the symbol of conjunction generalized to $\overline{q}$ conjuncts.

Proof. The proof will be carried on by induction on $\bar{q}$.

(I) Assume that $\bar{q}=1$ and observe that for any $t\in N$

(1) $\vdash (v_{s(t)})A_t^0(v_{s(t)})\to(v)A(v)$.

From (1) it follows that for any $t\in N$

(2) $\vdash (Ev_{s(t)})A_t^0(v_{s(t)})\to(v)A(v)$.

From (2) we infer for $t=1$, that

(3) the lemma holds for $\bar{q}=1$.

(II) Assume that $\bar{q}=t-1$, and that

(4) $\vdash (Ev_{s(1)})(Ev_{s(2)})\dots(Ev_{s(t-1)})\bigwedge_{q\leqslant t-1}[A_q^0(v_{s(q)})\to(v)A_q^0(v)]$.

It results from (2) and (4) that

(5) the lemma holds for $\bar{q}=t$.

Q. E. D.

LEMMA 4. *If* $X=\{B_1,B_2,\dots\}\subset S_1$ *is consistent, then there exists a normal valuation which is an element of the set*

$$H=\bigcap_k W(B_k)\cup\bigcap_p\Big[W\big(\neg(v)A_p(v)\big)\cup\bigcap_r W\big(A_p(c_r)\big)\Big]\cap$$
$$\cap\bigcap_q\Big[W\big((v)A_q(v)\big)\cup\bigcup_s W\big(\neg A_q(c_s)\big)\Big],$$

where the indices k, p, q, r, s *range over the set* N, *and the formulas of the form* $A_n(v)$ *are in the set* (+) *where defined above.*

Proof. Assume that

(1) $X\in Con$ and $X\subset S_1$

and suppose indirectly, that

(2) $H=\emptyset$.

In the definition of H replace the sum $\bigcup_s W\big(\neg A_q(c_s)\big)$ by one of its component, viz., by $W\big(\neg A_q(c_{s(q)})\big)$ where s is the function defined immediately before lemma 3. Therefrom it results by virtue of (2), that

(3) $$H'=\bigcap_k W(B_k)\cup\bigcap_{p,r}\Big[W\big(\neg(v)A_p(v)\big)\cup W\big(A_p(c_r)\big)\Big]\cap$$
$$\cap\bigcap_q\Big[W\big((v)A_q(v)\big)\cup W\big(\neg A_q(c_{s(q)})\big)\Big]=\emptyset.$$

From (3) and from the compactness of the space $\overline{W}^+$ it follows that there exist natural numbers $\bar{k}$, $\bar{p}$, $\bar{q}$, $\bar{r}$ which are dependent on $s(q)$ and such that

(4) $$H''=\bigcap_{k\leqslant\bar{k}}[W(B_k)\cap\bigcap_{\substack{p\leqslant\bar{p}\\ r\leqslant\bar{r}}}\Big[W\big(\neg(v)A_p(v)\big)\cup W\big(A_p(c_r)\big)\Big]\cap$$
$$\cap\bigcap_{q\leqslant\bar{q}}\Big[W\big((v)A_q(v)\big)\cup W\big(\neg A_q(c_{s(q)})\big)\Big]=\emptyset.$$

From (4) and from the lemmas 1 and 2, it follows that the formula

(5) $$\bigwedge_{k\leqslant\bar{k}} B_k \wedge \bigwedge_{\substack{p\leqslant\bar{p}\\ r\leqslant\bar{r}}} [(v)A_p(v)\to A_p(c_r)] \wedge \bigwedge_{q\leqslant\bar{q}} [A_q(c_{s(q)})\to (v)A_q(v)]$$

which is an element of S^+ is refutable.

By contraposition it results from (5) that

(6) $$\vdash \bigwedge_{\substack{p\leqslant\bar{p}\\ r\leqslant\bar{r}}} [(v)A_p(v)\to A_p(c_r)] \to \{\bigwedge_{q\leqslant\bar{q}} [A_q(c_{s(q)})\to (v)A_q(v)] \to \neg (\bigwedge_{k\leqslant\bar{k}} B_k)\}\,.$$

To every formula of the forms $(v)A_p(v)$ and $A_p(c_r)$ there is now assigned a formula of the form $A_p^0(v)$ and $A_p(v_r)$, respectively. From the notations adopted before lemma 3 it follows now that

(7) $$\vdash (v)A_p^0(v)\to A_p^0(v_r)\,.$$

By substitution and detachment, (6) and (7) yield that

(8) $$\vdash \bigwedge_{q\leqslant\bar{q}} [A_q^0(v_{s(q)})\to (v)A_q^0(v)] \to \neg (\bigwedge_{k\leqslant\bar{k}} B_k)\,.$$

From (8) by adjoining existential quantifiers in the antecedent, it follows on the basis of lemma 3 that

(9) $$\vdash \neg (\bigwedge_{q\leqslant\bar{q}} B_k)\,.$$

From (9) it results that

(10) $X \notin Con$

which contradicts (1).

From (10) it follows in spite of (2) that

(11) there exists a normal valuation which belongs to H.

Q. E. D.

Every normal valuation $\bar{w}$ determines a certain model $\mathfrak{M}_{\bar{w}} = \langle U, P, Q\rangle$ of the language L, defined by the following conditions:

(i) $U = C$,

(ii) $c_i \in P$ iff $\bar{w}P(c_i) = 1$,

(iii) $\langle c_i, c_j\rangle \in Q$ iff $\bar{w}Q(c_i, c_j) = 1$.

At the same time there holds the following:

LEMMA 5. *For any* $A \in S_1$

$$\mathfrak{M}_{\bar{w}} \models A \quad iff \quad \bar{w}A = 1\,.$$

Proof. For atomic formulas the thesis of the lemma is true by virtue of the conditions (ii)–(iii) above. The inductive step in passing from formulas B, D to formulas $\neg B$, $B\wedge D$ and $(x)B$ is immediate in view of the fact that $\bar{w}$ is the normal valuation.

Q. E. D.

Let us pass now to the proof of theorem 4. From the fact $S_1 \supset X \in Con$ it follows, according to the lemma 4, that there exists a normal valuation $\overline{w}_0 \in H$. In view of lemma 5, from the fact that $A \in X$, it results that $\mathfrak{M}_{\overline{w}_0} \models A$, i.e., $\mathfrak{M}_{\overline{w}_0}$ is a model for X.

IV. The method of A. Robinson of proving extended completeness theorem for *LPC* without identity

The method of A. Robinson was demonstrated in his paper [40] in 1961. Some slight modifications of this method are to be found in the paper [41], dated 1962 and in the monograph [42] published in 1963. The method itself amounts to a certain modification of Gödel's method. The problem of constructing of the model is reduced, however, on the grounds of Robinson's method, to the problem of completeness of the classical propositional calculus. This reduction is done, among others, by means of eliminating quantifiers according to the method of Skolem-Herbrand. It should be stressed that A. Robinson makes no reference to the familiar completeness theorem for the classical propositional calculus in extended sense which states that if every finite subset of a set of formulas is satisfiable in the two-element Boolean algebra, then all this set is satisfiable in this algebra. He gives instead the lemma fundamental for this method, which is referred to as the lemma on valuations. From this lemma there results, however, the completeness theorem for the classical propositional calculus (cf. the considerations following lemma 2 below).

A. Robinson's lemma on valuations is interesting due to its connections with the other important theorems. Namely, corollary 1 following from this lemma is a particular case of the familiar theorem of R. Rado. Corollary 2 will allow for giving a definition of the reduced ultraproduct of a family of models, while this definition is equivalent to that of J. Łoś quoted in Chapter IX. It will also be possible by virtue of this corollary to demonstrate a somewhat modified proof of the lemma on ultraproducts of J. Łoś (lemma 2 in Chapter IX).

1. A. Robinson's lemma on valuations

The name of partial valuation of arbitrary set Z will be given to every function w, such that there exists a set $W \subset Z$, such that $w\colon W \to \{0, 1\}$. If the domain $\mathfrak{D}w$ of the valuation w is equal to Z, then the partial valuation w will be called total valuation of the set Z.

By $w \upharpoonright Z'$ we shall symbolize the restriction of w to the subset $Z' \subset Z$.

The family $\mathfrak{F} \subset 2^I$ will be called a net on the non-empty set I provided that

(i) $\mathfrak{F} \neq \emptyset$,

(ii) $\emptyset \notin \mathfrak{F}$,

(iii) if $F_1 \in \mathfrak{F}$, and $F_2 \in \mathfrak{F}$, then there exists $F_3 \in \mathfrak{F}$, such that $F_1 \cap F_2 \supset F_3$.

The family $\mathfrak{F}$ satisfying apart from the conditions (i) and (ii), also the condition:

(iv) $F_1 \in \mathfrak{F}$ and $F_2 \in \mathfrak{F}$ iff $F_1 \cap F_2 \in \mathfrak{F}$

will be called a filter on the set I.

The filter $\mathfrak{F}$ which satisfies the condition:

(v) $F_1 \in \mathfrak{F}$ iff $I - F_1 \in \mathfrak{F}$

will be called an ultrafilter on the set I.

LEMMA 1 (*on valuations*). *If*

(1) *$\{w_\nu : \nu \in I\}$ is the set of partial valuations of the set Z,*

(2) *$\mathfrak{F}$ is a net on the set I,*

(3) *for any $Z' \in FinZ$, and for any $F \in \mathfrak{F}$, there exists $\nu \in F$, such that $Z' \subset \mathfrak{D}w_\nu$,*

then

(4) *there exists a total valuation f of the set Z, such that for every $Z' \in FinZ$, and for every $F \in \mathfrak{F}$, there exists $\nu \in F$, such*

$$Z' \subset \mathfrak{D}w \quad \textit{and} \quad f \upharpoonright Z' = w_\nu \upharpoonright Z'.$$

Proof. Cf. [40].

COROLLARY 1. *If*

(1) *$\{w_\nu : \nu \in I\}$ is the set of partial valuations of the set Z,*

(2) *for any $Z' \in FinZ$ there exists $\nu \in I$ such that $Z' \subset \mathfrak{D}w_\nu$*

then

(3) *there exists a total valuation f of the set Z such for any $Z' \in FinZ$ there exists $\nu \in I$ satisfying the condition:*

$$Z' \subset \mathfrak{D}w_\nu \quad \textit{and} \quad f \upharpoonright Z' = w_\nu \upharpoonright Z'.$$

Proof. Corollary 1 is an immediate consequence of lemma 1, where in place of $\mathfrak{F}$ there was put the net $\{I\}$.

Q. E. D.

COROLLARY 2. *If*

(1) *$\{w_\nu : \nu \in I\}$ is a set of partial valuations of the set Z,*

(2) *$\mathfrak{F}$ is an ultrafilter on I,*

(3) *for any* $Z' \in FinZ$ *and for any* $F \in \mathfrak{F}$ *there exists* $\nu \in F$ *such that* $Z' \subset \mathfrak{D}w_\nu$,

then

(4) *there exists exactly one total valuation* f *of the set* Z *such that for any* $Z' \in FinZ$ *and for any* $F \in \mathfrak{F}$ *there exists* $\nu \in F$ *for which*

$$Z' \subset \mathfrak{D}w \quad \textit{and} \quad f \upharpoonright Z' = w_\nu \upharpoonright Z' .$$

Proof. Assume (1)–(3). In view of lemma 1 we deduce that (5) there exists a total valuation f which satisfies the conditions formulated in (4).

Let $z \in Z$. Let us accept the following notations:

(6) $I_0(z) = \{\nu \in I\colon\ z \in \mathfrak{D}w_\nu \quad \text{and} \quad w_\nu(z) = 0\}$.

(7) $I_1(z) = \{\nu \in I\colon\ z \in \mathfrak{D}w_\nu \quad \text{and} \quad w_\nu(z) = 1\}$.

(8) $I(z) = \{\nu \in I\colon\ z \notin \mathfrak{D}w_\nu\}$.

From the definitions (6)–(8) it follows that

(9) the sets $I_0(z)$, $I_1(z)$ and $I(z)$ are disjoint and their sum equals to I.

From (2) and (9) we obtain that

(10) exactly one of the sets defined in (6)–(8) belongs to $\mathfrak{F}$.

We shall demonstrate that

(11) $I(z) \notin \mathfrak{F}$.

Aiming at that, let us assume that

(11.1) $I(z) \in \mathfrak{F}$.

From (11.1) and from (3), putting $Z' = \{z\}$ and $F = I(z)$, we infer that

(11.2) there exists $\nu_0 \in I(z)$ such that $z \in \mathfrak{D}w_{\nu_0}$.

However, from (8) it follows that we cannot accept (11.2). Thus there holds (11).

We shall demonstrate now that

(12) $I_1(z) \in \mathfrak{F}$ iff $f(z) = 1$.

Assume, aiming at that, that

(12.1) $I_1(z) \in \mathfrak{F}$.

From this and from (5) it follows that

(12.2) there exists $\nu_1 \in I_1(z)$ such that $z \in \mathfrak{D}w_{\nu_1}$ and $f(z) = w_{\nu_1}(z)$.

From (7) and (12.2) we infer that

(12.3) $f(z) = 1$.

Suppose now, conversely, that there holds (12.3), and indirectly assume that

(12.4) $I_1(z) \notin \mathfrak{F}$.

From (10), (11) and (12.4) it follows that

(12.5) $I_0(z) \in \mathfrak{F}$.

We obtain now from (5), (6) and (12.5) that

(12.6) there exists $\nu_0 \in I_0(z)$ such that $z \in \mathfrak{D}w_{\nu_0}$ and $f(z) = w_{\nu_0}(z) = 0$.

Observe now the contradiction between (12.6) and the assumption (12.3); thus

(12.7) $I_1(z) \in \mathfrak{F}$.

Analogously we demonstrate that

(13) $I_0(z) \in \mathfrak{F}$ iff $f(z) = 0$.

The uniqueness of the condition (4) results from (12) and (13).

Q. E. D.

With accepting in corollary 2 an assumption stronger than (1), which would state that $\{w_\nu : \nu \in I\}$ is the set of total valuations of the set Z, the proof becomes simpler in the case of the step (11). Namely, $I(z)$ is then the empty set and hence there immediately results that $I(z) \notin \mathfrak{F}$.

If X is an arbitrary set of formulas, then by its Skolem form, in symbols $skl(X)$, we shall understand the set of sentences thus obtained:

(i) We form a set Y of all sentences such that each of its element is a normal form of exactly one element from X.

(ii) Each of the formulas in the set X has its normal form in Y.

(iii) $skl(X) = Skl(Y)$, where $Skl(Y)$ is the set of Skolem forms which are obtained from the set Y by eliminating existential quantifiers in its elements.

By $Oskl(X)$ we shall denote the set of all open formulas obtained from the set $skl(X)$ by erasing all universal quantifiers in all the formulas in the set $skl(X)$.

Assume that

(+) $X_0 \in Con$.

Assumption (+) will be valid on the subsequent considerations in the present chapter. All construction below will be related to the set X_0.

By T we shall symbolize the smallest set of all terms satisfying the conditions:

(i) Every individual constant being the element of $skl(X_0)$ is also an element of T. In case no formula in X_0 has its normal form beginning with the existential quantifier, we include to the set T any individual constant.

(ii) If $t_1, \ldots, t_n \in T$ and F is an n-ary function symbol introduced to the set $skl(X_0)$, then $F(t_1, \ldots, t_n) \in T$.

By Y_0 we shall symbolize the set of all sentences which will be obtained from the set $Oskl(X_0)$ by substituting arbitrary terms from T for all free variables in his elements; at the same time we perform these substitutions in all possible manners.

By At_0 we shall symbolize the set of all atomic sentences constructed by means of predicate symbols which occur in the set X_0 and of the terms of the set T. Since in our language there occur only two predicate symbols P and Q, then let us assume that they occur in the set X_0. This assumption will allow us to avoid introducing additional symbols.

Let $h\colon At_0 \to Var$ be a bijection. By $\bar{h}$ we shall symbolize an isomorphic extension of h to the set Y_0 (cf. the considerations in § 5 in Chapter I).

LEMMA 2. *The set* $H = \bar{h}(Y_0)$ *is consistent in the classical propositional calculus.*

Proof. From the definition of H it results that H is a set of formulas in the language of the propositional calculus.

Assume that

(1) $H \notin Con$.

From (1) it follows that

(2) there exists $H_f \in FinH$ such that $H_f \notin Con$.

It results from (2) that

(3) $\bar{h}^{-1}(H_f) \notin Con$.

From (3) it follows that

(4) $Y_0 \notin Con$.

From the definition of the set Y_0 and from the lemma stating that if $X \in Con$, then $skl(X) \in Con$, it follows that

(5) $X_0 \notin Con$

which contradicts the assumption (+).

Q. E. D.

Let $Var^+ \in FinVar$. By H^+ we shall symbolize the set of all these sentences of the set H which are formed by means of the variables of the set Var^+. Neglecting the difference between tautologically equivalent formulas we may state that $H^+ \in FinH$.

In viev of the lemma 2 we infer that $H^+ \in Con$. Then there exists such a valuation $w\colon Var^+ \to \{0, 1\}$ that for any $A \in H^+$ $\bar{w}A = 1$, where $\bar{w}$ is Boolean valuation, of course.

By $\{w_\nu\colon \nu \in I\}$ we shall symbolize the set of all partial valuations of the set Var defined like above.

In view of corollary 1, we obtain that there exists a total valuation f of Var such that for every $Var' \in FinVar$ there exists $\nu \in I$ such that $Var' \subset \mathfrak{D}w_\nu$ and $f \upharpoonright Var' = w_\nu \upharpoonright Var'$.

LEMMA 3. *For every* $A \in H$ *the equality* $\bar{f}A = 1$ *holds.*

Proof. Let $V \in FinVar$ be the set of all variables of which A is composed. There exists then $\nu \in I$ such that $V \subset \mathfrak{D}w_\nu$ and $w_\nu \upharpoonright V = f \upharpoonright V$. Hence we infer that $\overline{w}_\nu \upharpoonright H^+ = \bar{f} \upharpoonright H^+$, where $H^+ \subset H$ is an arbitrary set of sentences which are composed exlusively of the variables in V. Since $A \in H^+$, then $\bar{f}A = \overline{w}A = 1$.

Q. E. D.

Let us pass now to the construction of the model for the language enriched with the set of individual constants $\{c_i : i \in J_1\}$, and the set of function symbols $\{F_i : i \in J_2\}$, which are introduced to the language L by applying Skolem's procedure of eliminating of existential quantifiers to the consistent set X_0.

The model $\mathfrak{M}^+ = \langle U, \{s_i : i \in J_1\}, P, Q, \{f_i : i \in J_2\}\rangle$ is defined by the conditions:

(i) $U = T$,

(ii) $s_i = c_i$, for every $i \in J_1$,

(iii) $t \in P$ iff $\bar{f}\bar{h}P(t) = 1$,

(iv) $\langle t_1, t_2\rangle \in Q$ iff $\bar{f}\bar{h}Q(t_1, t_2) = 1$,

(v) $f_i(t_1, \ldots, t_{\gamma(i)}) = F_i(t_1, \ldots, t_{\gamma(i)})$.

It is easy now to conclude from the above considerations the following:

LEMMA 4. *For any* $A \in Y_0$

$$\mathfrak{M}^+ \models A \quad \textit{iff} \quad \bar{f}\bar{h}A = 1 .$$

From lemma 4 there follows, in turn, the following:

LEMMA 5.

(a) *For any* $A(x_1, \ldots, x_n) \in Oskl(X_0)$ *and for any* $t_1, \ldots, t_n \in T$

$$\mathfrak{M}^+ \models A(x_1, \ldots, x_n)[t_1, \ldots, t_n] \quad \textit{iff} \quad \bar{f}\bar{h}A(t_1, \ldots, t_n) = 1 .$$

(b) *For any* $A \in skl(X_0)$, $\mathfrak{M}^+ \models A$.

We shall still formulate a quite general lemma which has a simple proof based on the definition of satisfaction and on the definition of the Skolem form; the proof is omitted here.

LEMMA 6. *If*

(1) $\mathfrak{M}$ *is an arbitrary model of the language* L,

(2) $A \in S_1$,

(3) *$\{s_i\colon i \in J_1\}$ and $\{F_i\colon i \in J_2\}$ are the sets of all individual and function constants introduced to the language L by applying Skolem's procedure of eliminating of existential quantifiers to A,*

then the following equivalence holds:

(4) *$\mathfrak{M} \models A$ iff there exist sets, $\{s_i\colon i \in J_1\}$ and $\{f_i\colon i \in J_2\}$ such that s_i and f_i are interpretations of c_i and F_i in the model, such that $\langle \mathfrak{M}, \{s_i\colon i \in J_1\}, \{f_i\colon i \in J_2\}\rangle \models skl(\{A\})$.*

Let us pass now to the conclusive fragment of the proof of theorem 3 by means of Robinson's method. From lemma 5(a) it follows on the basis of lemma 6 that the model $\mathfrak{M}_1^+$ which results from the model $\mathfrak{M}^+$ by erasing all the functions f_i and s_i, verifies the set X_1, where X_1 being composed exactly of the normal forms of all formulas in X_0. Hence it follows that $\mathfrak{M}_1^+$ is the model of X_0.

V. Henkin's method of proving the extended completeness theorem and some modifications of this method

1. Henkin's method

The method of L. Henkin published in 1949 in [13] based on one of the most elegant ways of construing models. It does not require the extensive formal elaboration of the *LPC*. It utilizes only some metatheorems pertaining to the properties of consistence, syntactical completeness as well as the deduction theorem for the *LPC*.

We shall make now some remarks which are to clear up the intuitions leading to Henkin's construction.

Assume that $Y \in Syst \cap Con \cap Com$. The set Y expressed in the language L with identity determines a certain model of this language. Namely, it is the model $\mathfrak{M} = \langle U, P, Q, Id\rangle$ defined by means of the following conditions:

(i) $U = \{v_i\colon i \in \omega\}/_{\sim}$,

(ii) $v_i^{\sim} \in P$ iff $P(v_i) \in Y$,

(iii) $\langle v_i^{\sim}, v_j^{\sim}\rangle \in Q$ iff $Q(v_i, v_j) \in Y$,

(iv) $\langle v_i^{\sim}, v_j^{\sim}\rangle \in Id$ iff $v_i \doteq v_j \in Y$.

The relation $\sim$ is defined in the set $\{v_i\colon i \in \omega\}$ by the condition: $v_i \sim v_j$ iff $v_i \doteq v_j \in Y$, and $v_i^{\sim} = \{v_j\colon v_i \sim v_j\}$. The relation $\sim$ is a congruence and it guarantees that Id is the identity in U.

The assumptions that are satisfied by Y function as follows. In view of the fact that $Y \in Syst$, we write, e.g., $P(v_i) \in Y$ instead of writing $Y \vdash P(v_i)$ which would have the same effect. In view of the fact that

$Y \in Con$, it follows that the relations P, Q and Id may be defined without a contradiction arising in the metasystem. In view of the fact that $Y \in Com$, it follows that with respect to any element and any pair of them, the question may be decided whether they belong to the respective relation or not, regarding the set Y.

The above observations will not suffice yet to state that the described model $\mathfrak{M}$ of the language L is the model of the set Y. It may happen that the set Y includes the formula $(Ev_i)A(v_i)$, and each of the formulas $\neg A(v_j)$. In this case there will be not enough elements in the domain U of the model $\mathfrak{M}$ and for the relations in the model it will be impossible to „hold" in such a way as it is described by the elements of the set Y. To overcome this difficulty the set $\{v_i\colon i \in \omega\}$ must be enriched with sufficiently many new individual variables (cf., e.g., [22], p. 51–56). Since we aim at a faithful reconstruction of Henkin's ideas we shall not then extend the set $\{v_i\colon i \in \omega\}$, but rather enrich the language with the set of individual constants. As an immediate consequence which does not at all diminish the generality of consideration, we have the construction of a model for the set of sentences. We shall pass to describe the construction and later we shall demonstrate that it is capable of being realized.

Let $C = \{c_i\colon i \in \omega\}$ be an infinite set of individual constants. By $L(C)$ we shall symbolize the language obtained from the language L by adding to it the set C. Observe that it is inessential to assume the set C to be denumerable. It will suffice to assume that the cardinality of C is infinite. We shall make use of this fact while comparing Henkin's method with that of ultraproducts.

Any set of sentences in the language $L(C)$, such that $Y \in Syst \cap Con \cap Com$, determines the model

$$\mathfrak{M}(Y, C) = \langle U, \{s_i\colon i \in \omega\}, P, Q, Id\rangle$$

of the language $L(C)$ in which

(i) $U = C/_{\sim}$,

(ii) $s_i = c_i^{\sim}$,

(iii) $c_i^{\sim} \in P$ iff $P(c_i) \in Y$,

(iv) $\langle c_i^{\sim}, c_j^{\sim}\rangle \in Q$ iff $Q(c_i, c_j) \in Y$,

(v) $\langle c_i^{\sim}, c_j^{\sim}\rangle \in Id$ iff $c_i \doteq c_j \in Y$,

where the relation $\sim$ is defined in C by means of the condition: $c_i \sim c_j$ iff $c_i \doteq c_j \in Y$ and $c_i^{\sim} = \{c_j \in C\colon c_i \sim c_j\}$.

We say that the set of sentences Y satisfies the condition ($\mathfrak{H}$) in the set of constants C, provided that for every existential sentence $(Ex)A(x)$, from the fact that $(Ex)A(x) \in Y$, it follows that there exists an individual constant $c \in C$ such that $A(x/c) \in Y$.

We shall now give the lemma fundamental for the method of Henkin.

LEMMA 1. *If* Y *is the set of sentences in the language* $L(C)$ *such that*

(1) $Y \in Syst \cap Con \cap Com$,

(2) Y *satisfies the condition* $(\mathfrak{H})$ *in the set* C,

then for any sentence A

(3) $\mathfrak{M}(Y, C) \models A$ *iff* $A \in Y$.

Proof. Assume (1) and (2). For the atomic sentences the equivalence (3) holds by virtue of the definition of the model (Y, C). Assume then, for the sake of induction, that the equivalence (3) holds for the sentences A_1 and A_2, i.e., that

(4) $\mathfrak{M}(Y, C) \models A_1$ iff $A_1 \in Y$

and

(5) $\mathfrak{M}(Y, C) \models A_2$ iff $A_2 \in Y$.

From the definition of satisfaction it follows that the condition

(6) $\mathfrak{M}(Y, C) \models \neg A_1$

is equivalent to the condition

(7) non $\mathfrak{M}(Y, C) \models A_1$.

We infer from (4) that (7) is equivalent to the fact that

(8) $A_1 \notin Y$.

In turn, it follows from (1) that (8) equals to the fact that

(9) $\neg A_1 \in Y$.

Hence and from (5) we deduce that

(10) the condition (3) holds for the formulas of the form $\neg A_1$.

In an analogous way we prove basing on the assumptions (1), (4) and (5) that

(11) the condition (3) holds for the formulas of the form $A_1 \wedge A_2$.

Assume now that $(Ex)A(x)$ is a sentence and that

(12) the condition (3) holds for any formula of the form $A(c)$.

Assume, moreover, that

(13) $\mathfrak{M}(Y, C) \models (Ex)A(x)$

and suppose indirectly that

(14) $(Ex)A(x) \notin Y$.

Hence and from (1) it results that

(15) $(x)\neg A(x) \in Y$.

This, together with (1) and with the axiom (A14), yields that for any $c \in C$

(16) $\neg A(c) \in Y$.

Hence, by the inductive assumption (12) we infer that for any $c \in C$

(17) $\mathfrak{M}(Y, C) \models \neg A(c)$.

Hence, in view of the definition of satisfaction it follows that for any $c \in C$

(18) non $\mathfrak{M}(Y, C) \models A(c)$.

But from (13) it follows that there exists such $c \in C$ that

(19) $\mathfrak{M}(Y, C) \models A(x)[c^{\sim}]$.

This yields in view of the definition of satisfaction, that there exists $c \in C$ such that

(20) $\mathfrak{M}(Y, C) \models A(c)$.

Hence and from (18) we deduce that (14) cannot be accepted, i.e.,

(21) $(Ex)A(x) \in Y$.

Assume now that (21) holds. Then, it follows from (2) that there exists $c \in C$ such that

(22) $A(c) \in Y$.

This yields, together with (12), that there exists $c \in C$ such that

(23) $\mathfrak{M}(Y, C) \models A(c)$.

Hence, by the definition of satisfaction it results that

(24) $\mathfrak{M}(Y, C) \models (Ex)A(x)$.

From (13), (21) and (24) we infer that

(25) the equivalence (3) holds for the formulas of the form $(Ex)A(x)$.

Q. E. D.

Below we quote a few of the familiar lemmas. They are recalled here to make the remaining proofs clearer.

LEMMA 2 (the deduction theorem). *If A and B are sentences, then from the fact that $X \cup \{A\} \vdash B$ it follows that $X \vdash A \rightarrow B$.*

Proof. Cf., e.g., [28], p. 61.

LEMMA 3. *If*

(1) *the individual constant c does not occur in the formulas of the set X,*

(2) *$A(v)$ results from $A(c)$ by substituting the variable v in place of c in the formula $A(c)$ and v is not, at the same time, a bound variable anywhere in $A(v)$*

and

(3) $X \vdash A(c)$

then

(4) $X \vdash A(v)$.

Proof. As regards the proof of lemma 3 which informs that *LPC* does not favour any individual constant cf., e.g., [10], p. 196.

LEMMA 4. *If*

(1) $X \in Con$,

(2) *the constant c does not appear among the elements of the set X,*

and

(3) $(Ex)A(x) \in X$

then

(4) $X \cup \{A(x/c)\} \in Con$.

Proof. Assume (1)–(3). Suppose indirectly that

(5) $X \cup \{A(c)\} \notin Con$.

Hence, in view of the definition of Con, it follows that

(6) $X \cup \{A(c)\} \rightarrow \neg A(c)$.

This and lemma 2 yield that

(7) $X \vdash A(c) \rightarrow \neg A(c)$.

Hence, in view of the thesis $(B \rightarrow \neg B) \rightarrow \neg B$, we infer that

(8) $X \vdash \neg A(c)$.

This and (2) yield in view of lemma 3, that

(9) $X \vdash \neg A(x)$.

It follows therefrom by the rule of generalization that

(10) $X \vdash (x) \neg A(x)$.

Hence, by De Morgan Law we obtain that

(11) $X \vdash \neg (Ex)A(x)$.

This and (3) yield that

(12) $X \notin Con$

which contradicts (1). Thus the supposition (5) cannot be accepted.

Q. E. D.

For the sake of the further considerations we shall introduce the following notations.

For any $i \in N$, let C_i be an infinite one-to-one sequence such that

$$C_i = \langle c_j^i \colon j \in N \rangle$$

and

$$C = \bigcup_{i \in \omega} C_i .$$

By L_n we shall symbolize the language which is an extension of the language L, denoted now by L_0, obtained by enriching the former with the set of individual constants $\bigcup_{i=1}^{n} C_i$.

By $S_{(0)}$ we shall symbolize the set of all sentences in the language L_0, and by $S_{(n)}$ the set of all sentences in the language L_n will be symbolized.

LEMMA 5. *If*

(1) $X \in Con$,

(2) $X \subset S_{(n)}$,

then there exists the set Y *such that*

(3) $X \subset Y$,

(4) $Y \subset S_{(n+1)}$,

(5) $Y \in Con$

and

(6) Y *satisfies the condition* ($\mathfrak{H}$) *regarding* X *in the set* $\bigcup_{i=1}^{n+1} C_i$, *i.e., for any sentence* $(Ex)A(x) \in X$ *there exists* $c \in \bigcup_{i=1}^{n+1} C_i$ *such that* $A(c) \in Y$.

Proof. Let $\langle (Ex_{k_j})A_j(x_{k_j}) \colon j \in N \rangle$ be the sequence of all existential sentences included in the set X. Let us put

(7) $Y = X \cup \{A_j(c_j^{n+1}) \colon j \in N\}$.

From (1) and from the definition of the sets C_i we infer on the grounds of the lemma 4, that the condition (5) is satisfied. The lines (3), (4) and (6) are immediate consequences of (7)

Q. E. D.

LEMMA 6. *If*

(1) $X \subset S_{(0)}$,

(2) $X \in Con$,

then there exists the set Y *such that*

(3) Y *is the set of sentences in the language* $L(C)$,

(4) $X \subset Y$,

(5) $Y \in Syst \cap Con \cap Com$

and

(6) Y *satisfies the condition* ($\mathfrak{H}$) *in* C.

Proof. Assume (1) and (2). We shall construct the sequence of sets

(7) $X_0, X_0^+, X_1, X_1^+, X_2, X_2^+, \ldots$

defined by the conditions:

(7.1) $X_0 = X_0^+ = X$

(7.2) X_n, where $n \geqslant 1$, is the consistent extension of X_{n-1}^+, which exist by virtue of lemma 5.

(7.3) X_n^+, where $n \geqslant 1$, is a consistent and complete Lindenbaum supersystem for the set X_n, and moreover, both X_n and X_n^+ are expressed in the language L_n.

Let us put that

(8) $Y = \bigcup_{i \in \omega} (X_i \cup X_i^+)$.

Observe that in view of (7) and (8) we have that

(9) $Y = \bigcup_{i \in \omega} X_i^+$.

From the definition (8) it follows that

(10) Y satisfies the conditions (3) and (4).

From (9), in view of the theorem on the sum of consistent and complete systems, it follows that

(11) Y satisfies the condition (5).

We shall now prove (6). Aiming at achieving that, let us assume that

(12) $(Ex)A(x) \in Y$.

Hence and from (9) we infer that there exists such n that

(13) $(Ex)A(x) \in X_{n-1}^+$.

This and (7.2) yield that there exists $c \in C$ such that

(14) $A(c) \in X_n$.

We deduce from (8) and (14) that

(15) $A(c) \in Y$.

This, together with (12), yields that

(16) Y has the property (6).

Q. E. D.

The extended completeness theorem 4 is the immediate conclusion from lemmas 1 and 6.

2. Hasenjaeger's simplifications

G. Hasenjaeger gave in 1953 in [12] a simpler proof of the crucial lemma 6. This we shall now briefly present.

Assume that $X \subset S_{(0)}$ and that $X \in Con$. Let

$A_1(x_{k_1}), A_2(x_{k_2}), \ldots, A_j(x_{k_j}), \ldots$

be the sequence of all formulas (not only in the set X), expressed in the language $L(C)$ where there is exactly one free variable occurring in each of them; namely x_{k_j} is free in $A_j(x_{k_j})$.

Select now the sequence of the individual constants

$c_{i_1}, c_{i_2}, \ldots, c_{i_j}, \ldots$

belonging to the set C, and such that no term of the sequence $A_1(x_{k_1}), \ldots, A_j(x_{k_j})$ contains the term c_{i_j}.

Construct the sequence of the sets

$X_0, X_1, \ldots, X_n, \ldots$

defined by the conditions:

$X_0 = X\,,$

$X_n = X_{n-1} \cup \{(Ex_{k_j})A(x_{k_j}) \to A(c_{i_j})\}\,.$

Let us put

$$Y = \bigcup_{n \in \omega} X_n\,.$$

It is easy to check that for every n, $X_n \in Con$. Hence it follows that $Y \in Con$.

Let Y^+ be the consistent and complete supersystem of Lindenbaum of the set Y (cf. lemma 4 in Chapter I). It is to be easily seen that Y has the property $(\mathfrak{H})$ in the set C which completes the proof of lemma 6.

3. Another modification of Henkin's construction

A still further reaching simplification of the proof of lemma 6 was given by Henkin himself (cf. [48], p. 96). It may also be found in [10], p. 125. The simplification amounts to combining Lindenbaum's construction with that of Hasenjaeger.

Assume then that $X \subset S_{(0)}$, and that $X \in Con$. Let

$A_1, A_2, \ldots, A_n, \ldots$

be the sequence of all sentences in the language $L(C)$.

The sequence of sets

$X_0, X_1, \ldots, X_n, \ldots$

is defined by means of the conditions:

$X_0 = X$,

$X_{n+1} = X_n \cup \{A_n\}$, if $X_n \cup \{A_n\} \in Con$ and A_n does not begin with the existential quantifier.

$X_{n+1} = X_n \cup \{A_n \to B(c_j)\}$, if $X_n \cup \{A_n\} \in Con$, there exists $B(x_i)$ such that $A_n = (Ex_i)B(x_i)$ and c_j is the first constant in the sequence $\langle c_i\colon i \in \omega\rangle$ not occurring in $A_1, \ldots, A_n$.

$X_{n+1} = X_n$, if $X_n \cup \{A_n\} \notin Con$.

It is an easy task to check that $Y = \bigcup_{n\in\omega} X_n \in Syst \cap Con \cap Com$ and that Y has the property ($\mathfrak{H}$) in the set C.

VI. The method of Rasiowa-Sikorski of proving the extended completeness theorem

The method of H. Rasiowa and R. Sikorski belongs to the group of algebraic methods. It was published in 1951 (cf. [34] or [35]). It started the fruitful applications of algebraic methods in logic, and particularly the methods of Boolean algebras. Thus realized were the ideas of A. Lindenbaum which he formed in late twenties. Roughly speaking, it amounts to considering formalized languages as certain abstract algebras construed of formulas in these languages.

1. Auxiliary facts of the theory of Boolean algebras

Assume that $\mathfrak{B} = \langle B, \cup, \cap, -, \leqslant, 0, 1\rangle$ is a Boolean algebra where $\cup, \cap, -, \leqslant, 0, 1$ are, respectively, the symbols of join, meet, complement, ordering (defined by the equivalence: $a \leqslant b$ iff $a \cap b = a$), zero and unit of this algebra.

Let I be an arbitrary set of indices. Assume that there exists $\sup_{\leqslant}\{a_i \in B\colon i \in I\} = a$, i.e., there exists the sum

(+) $\quad a = \bigcup_{i\in I} a_i$.

We say that an ultrafilter $\mathfrak{F}$ in an algebra $\mathfrak{B}$ preserves the sum (+) provided the fact that $a \in \mathfrak{F}$ is equivalent to the existence of $i_0 \in I$ such that $a_{i_0} \in \mathfrak{F}$.

Fundamental for this method is the following lemma of Rasiowa-Sikorski which is sometimes referred to as Tarski's lemma (cf. [2], p. 21).

LEMMA 1. *If*

(1) $\mathfrak{B}$ *is a Boolean algebra,*

(2) $\{I_n\colon n \in N\}$ is a denumerable family of the sets of indices,

(3) $a_0, a_n, a_{n,i} \in B$,

(4) $a_n = \bigcup_{i \in I_n} a_{n,i}$

and

(5) $a_0 \neq 0$,

then in the algebra $\mathfrak{B}$ *there exists an ultrafilter* $\mathfrak{F}$, *which preserves the sums* (4) *and such that* $a_0 \in \mathfrak{F}$.

Proof. It is very easy to prove by induction that

(6) for every n there exist indices $i_1, \ldots, i_n$ such that $i_n \in I_n$, and $a_0 \cap (a_1 \cup -a_{1,i_1}) \cap (a_2 \cup -a_{2,i_2}) \cap \ldots \cap (a_n \cup -a_{n,i_n}) \neq 0$.

From (6) it follows that

(7) the set $\{a_0, a_1 \cup -a_{1,i_1}, a_2 \cup -a_{2,i_2}, \ldots\}$ has the finite intersection property.

In view of (7) and of the familiar lemma of theory of Boolean algebras it follows that

(8) there exists an ultrafilter $\mathfrak{F}$ such that $a_0 \in \mathfrak{F}$, and for every n: $a_n \cup -a_{n,i_n} \in \mathfrak{F}$.

From (8) we have that

(9.1) $a_n \in \mathfrak{F}$ and $-a_{n,i_n} \notin \mathfrak{F}$

or

(9.2) $a_n \in \mathfrak{F}$ and $-a_{n,i_n} \in \mathfrak{F}$

or

(9.3) $a_n \notin \mathfrak{F}$ and $-a_{n,i_n} \in \mathfrak{F}$.

It is easy to check basing upon the assumption (4) that in each of the cases (9.1)–(9.3) we have that

(10) $\mathfrak{F}$ preserves the sums defined in (4).

The thesis of the lemma follows from (8) and (10).

Q. E. D.

The algebraic proof of lemma 1 given above, is based on the suggestions of Tarski and is contained in Feferman's review [7]. Rasiowa and Sikorski presented a topological proof of lemma 1 making use of a modification of the familiar Bair's theorem, which was given by Čech in [4], and which states:

If Z is a compact Hausdorff space, then every subset W of the first category is boundary set, i.e., $Z-W$ is dense.

In both methods the assumption (2) plays an essential role. In case of Tarski's method the assumption may be omitted applying the principle of generalized induction instead. In case of Rasiowa-Sikorski's method, however, this can not be done in view of the quoted theorem which is not capable of appropriate generalization of Stone spaces.

2. Lindenbaum Q-algebra and the proof of theorem 3

For the further considerations contained in this chapter we shall assume that

(a) X is a consistent set of formulas of the language L with identity,

(b) $\approx$ is a binary relation defined in the set S of all formulas by the condition:

$A \approx B$ iff $X \vdash A \leftrightarrow B$

(c) $|A| = \{B \in S:\ A \approx B\}$

(d) $S/_{\approx} = \{|A|:\ A \in S\}$

(e) $|A| \cap |B| = |A \wedge B|$

(f) $-|A| = |\neg A|$

(g) $|A| = 1$ iff $X \vdash A$

(h) $|A| = 0$ iff $X \vdash \neg A$

(i) $|A| \leqslant |B|$ iff $X \vdash A \rightarrow B$.

By virtue of these assumptions the following lemmas 2 and 3 hold.

LEMMA 2.

(a) *Relation* $\approx$ *is a congruence in Lindenbaum algebra* $L = \langle S, \neg, \wedge \rangle$ *of all formulas;*

(b) *The quotient algebra* $L/_{\approx} = \langle S/_{\approx}, \cap, -, 0, 1 \rangle$ *is a non-degenerate Boolean algebra (i.e.,* $\overline{\overline{S/_{\approx}}} \geqslant 2$*).*

LEMMA 3. *If*

(1) $A(v_k)$ *is a formula where* v_k *is a free variable,*

(2) $A(v_k//v_p)$ *is a formula obtained from* $A(v_k)$ *by substituting first all bound occurrences of the variable* v_p *in* $A(v_k)$ *by the first variable of the sequence* $\langle v_i:\ i \in \omega \rangle$ *which does not occur in* $A(v_k)$, *and then by substituting all free occurrences of* v_k *by* v_p,

then in the algebra $L/_{\approx}$, *there holds the equality:*

$$|(Ev_k)A(v_k)| = \bigcup_{p \in \omega} |A(v_k//v_p)|,$$

which states that $|(Ev_k)A(v_k)|$ *is the least upper bound of the set* $\{|A(v_k//v_p)|: p \in \omega\}$ *in the algebra* $L/_\approx$ *with respect to the relation* $\leqslant$ *defined above in* (i).

Proof. Assume (1) and (2). From the axiom (A14) it follows that

(3) $X \vdash A(v_k//v_p) \to (Ev_k)A(v_k)$.

From (3) and from the definition of $\leqslant$ we deduce that

(4) $|A(v_k//v_p)| \leqslant |(Ev_k)A(v_k)|$.

Assume now that B is an arbitrary formula satisfying

(5) $|A(v_k//v_p)| \leqslant |B|$.

From (5) and from the definition of $\leqslant$ we obtain that

(6) $X \vdash A(v_k//v_p) \to B$.

Let us choose now q such that v_q does not occur in B. From (6) we shall obtain that

(7) $X \vdash (Ev_q)A(v_k//v_q) \to B$.

Axiom (A14) yields that

(8) $X \vdash A(v_k) \to (Ev_q)A(v_k//v_q)$.

From (8) in turn, we deduce by virtue of (2) that

(9) $X \vdash (Ev_k)A(v_k) \to (Ev_q)A(v_k//v_q)$.

From (7) and (9) it follows that

(10) $X \vdash (Ev_k)A(v_k) \to B$.

From (10) we infer that

(11) $|(Ev_k)A(v_k)| \leqslant |B|$.

Conditions (4), (5) and (11) yield

(12) $|(Ev_k)A(v_k)| = \bigcup_{p \in \omega} |A(v_k//v_p)|$.

Q. E. D.

Algebra $L/_\approx$ which has the property described in lemma 3 will be called the Lindenbaum Q-algebra determined by the consistent set X. In case $X = \emptyset$ algebra $L/_\approx$ is called the Lindenbaum Q-algebra of the *LPC*.

We shall pass now to proving theorem 3. Assume then, that

(1) $X \in Con$.

In view of lemma 2 we deduce from (1) that

(2) $L/_\approx$ is a non-degenerate Boolean algebra.

Let

(3) $\langle B_n : n \in N \rangle$ be the sequence of all existential formulas included in the set S, and of the form $B_n = (Ev_{k_n})A_n(v_{k_n})$.

From (3) and from lemma 3 it follows that

(4) $$|(Ev_{k_n})A_n(v_{k_n})| = \bigcup_{p_n \in \omega} |A_n(v_{k_n}//v_{p_n})| .$$

It is easy to see in view of (3) that the family of sums defined in (4) is denumerable. Thus it follows from lemma 1 that in algebra $L/_{\approx}$:

(5) there exists an ultrafilter $\mathfrak{F}$ which preserves the sums (4).

(6) The relation $\sim$ defined in the set $V = \{v_i : i \in \omega\}$ by the condition: $v_i \sim v_j$ iff $|v_i \doteq v_j| \in \mathfrak{F}$ is an equivalence.

From (5) and (6) we infer that we are able to define a model $\mathfrak{M}_{\mathfrak{F}} = \langle U, P, Q, Id\rangle$ with an absolute concept of identity, and in which

(i) $U = V/_{\sim}$,

(ii) $v_i^{\sim} \in P$ iff $|P(v_i)| \in \mathfrak{F}$,

(iii) $\langle v_i^{\sim}, v_j^{\sim}\rangle \in Q$ iff $|Q(v_i, v_j)| \in \mathfrak{F}$,

(iv) $\langle v_i^{\sim}, v_j^{\sim}\rangle \in Id$ iff $|v_i \doteq v_j| \in \mathfrak{F}$.

Let $\bar{u} \in U^{\omega}$, where $\bar{u} = \langle v_{i_0}^{\sim}, v_{i_1}^{\sim}, \ldots, v_{i_n}^{\sim}, \ldots\rangle$ and let $A(x_0, \ldots, x_n)$ be a formula where $x_0, \ldots, x_n$ are all free variables.

It is inductively proved now that

(7) $$\mathfrak{M}_{\mathfrak{F}} \models A(x_0, \ldots, x_n)[\bar{u}] \text{ iff } |A(v_{i_0}, \ldots, v_{i_n})| \in \mathfrak{F} .$$

The inductive step in passing from formulas A and B to the formula of the form $A \wedge B$ meets its justification in (7) with the help of the fact that $\mathfrak{F}$ if a filter; passing to the formula of the form $\neg A$ is justified by the fact that $\mathfrak{F}$ is an ultrafilter, whereas passing to the formula of the form $(Ev_i)A$ is motivated by the fact that $\mathfrak{F}$ is an ultrafilter which preserves the sums (4), and that in (4) these sums are determined for every existential formula.

Let us now pass to the concluding part of the proof of theorem 3. From the fact that $X \in Con$ it follows that for all $A \in X$, $|A| = 1$. Hence we deduce that $|A| \in \mathfrak{F}$. From the definition of the model $\mathfrak{M}_{\mathfrak{F}}$ and from the equivalence (7) we infer that $\mathfrak{M}_{\mathfrak{F}}$ is the model of the set X.

Observe, moreover, that in the proof of theorem 3, which was based on fundamental lemma 1, it would suffice to base it on a weaker version of this lemma. This weaker version states that if $\mathfrak{B}$ is a non-degenerate Boolean algebra and if there hold assumptions (2)–(4) of lemma 1, then in algebra $\mathfrak{B}$ there exists an ultrafilter that preserves the sums (4). Lemma 1 is formulated in such a way as to allow its application to the proof of theorem 1 on completeness in narrower sense. We shall demonstrate this proof now.

Assume that A_0 is not a thesis. Then in Lindenbaum Q-algebra of the LPC $|\neg A_0| \neq 0$. Thus in view of lemma 1 there exists in this algebra an ultrafilter $\mathfrak{F}$ such that $|\neg A_0| \in \mathfrak{F}$, and which preserves the sums (4) defined in the proof of theorem 3 above. The constructed model $\mathfrak{M}_{\mathfrak{F}}$ verifies the formula $\neg A_0$. Hence it follows that A_0 is not a tautology which completes the proof of theorem 1.

VII. Rieger's method of proving the extended completeness theorem

The algebraic method of L. Rieger has been published in 1950 in [37] and it is very closely related to the method of Rasiowa-Sikorski. Lindenbaum Q algebra is here treated as some Boolean $\Phi\sigma$ algebra. The concept of the Boolean $\Phi\sigma$-algebra is due to L. Rieger and it is a generalization of the concept σ-algebra (i.e., denumerably additive Boolean algebra).

The fundamental difficulty of mathematical character is, on the grounds of this method, to give the lemma on the existence of $\Phi\sigma$-ultrafilters; this lemma corresponds to the lemma of Rasiowa-Sikorski. Rieger deduces this lemma from the representation theorem for the denumerable Boolean $\Phi\sigma$-algebras with the denumerable family Φ of the marked sequences. The mentioned representating theorem, however, is obtained by L. Rieger by modifying the familiar Stone's method and basing on the theorem of Loomis (cf. [43], p. 117) as well as on the so-called method of cuts of MacNeille [25]. Thus, the mathematical apparatus utilized by L. Rieger is more complicated when compared with the other methods.

Observe still that L. Rieger in the paper [38] of 1951 gave a somewhat different method of proving the Gödel-Malcev's theorem. This he makes basing on the fact that Lindenbaum $\Phi\sigma$-algebra is representable by a subfield of the set-field of Borel subsets of Cantor discontinuum.

In the monograph [39] which is entirely devoted to algebraic methods in logic, L. Rieger gives a still different representation of the Lindenbaum $\Phi\sigma$-algebra. Namely, he demonstrates that it is isomorphic with a certain, what he calls, substitutively indexed Boolean algebra. From this fact there follows still another modification of the proof of the extended completeness theorem.

L. Rieger's scientific activity in logic was strongly connected with the Polish logical and mathematical circles. The results of his work [37] were presented by L. Rieger at the seminar of Professor A. Mostowski in Warsaw, during April 1950.

1. The concept of Boolean $\Phi\sigma$-algebra

Let Φ be a family of multiple sequences composed of the elements of some Boolean algebra $\mathfrak{B}$. These sequences will be symbolized by

$$\{a_{n_1, n_2, \ldots, n_k} : n_1, n_2, \ldots, n_k \in \omega\}$$

and

$$\{b_{m_1, m_2, \ldots, m_l}: m_1, m_2, \ldots, m_l \in \omega\}$$

and will be called marked sequences.

Boolean algebra $\mathfrak{B}$ will be called a Boolean $\Phi\sigma$-algebra provided that the set Φ of the marked sequences satisfies the following conditions:

(i) If $\{a_{n_1,\ldots,n_k}\} \in \Phi$, then $\{-a_{n_1,\ldots,n_k}\} \in \Phi$.

(ii) If $\{a_{n_1,\ldots,n_k}\} \in \Phi$ and $\{b_{m_1,\ldots,m_l}\} \in \Phi$, then $\{a_{n_1,\ldots,n_k} \cup b_{m_1,\ldots,m_l}\} \in \Phi$ and $\{a_{n_1,\ldots,n_k} \cap b_{m_1,\ldots,m_l}\} \in \Phi$.

(iii) If $\{a_{n_1,\ldots,n_k}\} \in \Phi$ and $0 \leqslant i_1 < \ldots < i_s \leqslant k$, then $\{a_{n_1,\ldots,n_{i_1},\ldots,n_{i_s},\ldots,n_k}: n_1, \ldots, n_{i_1-1}, n_{i_1+1}, \ldots, n_{i_s-1}, n_{i_s+1}, \ldots, n_k \in \omega\} \in \Phi$

(iv) If $\{a_{n_1,\ldots,n_k}\} \in \Phi$ and $0 \leqslant i_1 < \ldots < i_s \leqslant k$, then $\{a_{n_1,\ldots,n_{i_1-1},n,n_{i_1+1},\ldots,n_{i_s-1},n,n_{i_s+1},\ldots,n_k}\} \in \Phi$.

(v) For any $a \in \mathfrak{B}$ and for any $n_1, \ldots, n_k \in \omega$ if $a_{n_1,\ldots,n_k} = a$, then $\{a_{n_1,\ldots,n_k}\} \in \Phi$.

(vi) If $\{a_{n_1,\ldots,n_k}\} \in \Phi$, then there exist in algebra $\mathfrak{B}$ the elements a and b such that

$$a = \bigcup_{n_1} \bigcup_{n_2} \ldots \bigcup_{n_k} a_{n_1, n_2, \ldots, n_k} = \bigcup_{n_1,\ldots,n_k} a_{n_1,\ldots,n_k}$$

and

$$b = \bigcap_{n_1} \bigcap_{n_2} \ldots \bigcap_{n_k} a_{n_1, n_2, \ldots, n_k} = \bigcap_{n_1,\ldots,n_k} a_{n_1,\ldots,n_k}$$

(vii) If $\{a_{n_1,\ldots,n_k}\} \in \Phi$, $0 \leqslant j \leqslant k$

and

$$b_{n_1,\ldots,n_{j-1},n_{j+1},\ldots,n_k} = \bigcup_{n_j} a_{n_1,\ldots,n_j,\ldots,n_k}$$

and

$$c_{n_1,\ldots,n_{j-1},n_{j+1},\ldots,n_k} = \bigcap_{n_j} a_{n_1,\ldots,n_j,\ldots,n_k}$$

then

$$\{b_{n_1,\ldots,n_{j-1},n_{j+1},\ldots,n_k}: n_1, \ldots, n_{j-1}, n_{j+1}, \ldots, n_k \in \omega\} \in \Phi$$

and

$$\{c_{n_1,\ldots,n_{j-1},n_{j+1},\ldots,n_k}: n_1, \ldots, n_{j-1}, n_{j+1}, \ldots, n_k \in \omega\} \in \Phi .$$

The condition (iv) is the rule of forming diagonal sequences while condition (iii) is the rule of forming cylindric sequences. In the conditions (iii) and (vii) we have made an explicit statement of the ranges of indices.

The concept of Boolean $\Phi\sigma$-algebra is a generalization of the concept of denumerable additive Boolean algebra, sometimes referred to as Boolean σ-algebra. $\Phi\sigma$-algebra is the σ-algebra in the case when Φ is a family of all multiple sequences. In the case when Φ is a family of constant sequences every Boolean algebra is a $\Phi\sigma$-algebra.

A non-empty set $\mathfrak{F} \subsetneq \mathfrak{B}$ is said to be a $\Phi\sigma$-ultrafilter in the $\Phi\sigma$-algebra provided $\mathfrak{F}$ satisfies the conditions:

(i) If $b \in \mathfrak{F}$ and $b \leqslant a$, then $a \in \mathfrak{F}$.

(ii) If $\{a_{n_1,\dots,n_k}\} \in \Phi$ and $a_{n_1,\dots,n_k} \in \mathfrak{F}$, then $\bigcap\limits_{n_1,\dots,n_k} a_{n_1,\dots,n_k} \in \mathfrak{F}$, for every $n_1, \dots, n_k \in \omega$.

(iii) If $\{a_{n_1,\dots,n_k}\} \in \Phi$ and $\bigcup\limits_{n_1,\dots,n_k} a_{n_1,\dots,n_k} \in \mathfrak{F}$, then there exist indices $\bar{n}_1, \dots, \bar{n}_k$ such that $a_{\bar{n}_1,\dots,\bar{n}_k} \in \mathfrak{F}$.

If $\mathfrak{A}$ is a Boolean $\Phi_1\sigma$-algebra, then an isomorphism f of the algebra $\mathfrak{A}$ onto $\Phi_2\sigma$-algebra $\mathfrak{B}$ is a generalized σ-isomorphism if and only if for every $\{b_{m_1,\dots,m_l}\} \in \Phi_2$ there exists $\{a_{m_1,\dots,m_l}\} \in \Phi_1$ such that

$$f(a_{m_1,\dots,m_l}) = b_{m_1,\dots,m_l}$$

and such that there hold the following equalities:

$$\bigcap_{n_1,\dots,n_k} f(a_{n_1,\dots,n_k}) = f\Big(\bigcap_{n_1,\dots,n_k} a_{n_1,\dots,n_k}\Big)$$

and

$$\bigcup_{n_1,\dots,n_k} f(a_{n_1,\dots,n_k}) = f\Big(\bigcup_{n_1,\dots,n_k} a_{n_1,\dots,n_k}\Big)$$

The $\Phi_1\sigma$-algebra $\mathfrak{A}$ is a subalgebra of the $\Phi_2\sigma$-algebra $\mathfrak{B}$ provided that $\mathfrak{A}$ is a subalgebra of $\mathfrak{B}$ in the usual sense and if $\Phi_1 \subset \Phi_2$.

Let us quote now the lemmas which allow to give the proof of the representation theorem for Boolean $\Phi\sigma$-algebras.

LEMMA 1. *If $\mathfrak{B}$ is a Boolean $\Phi\sigma$-algebra then there exists a Boolean σ-algebra $\mathfrak{A}$ in which $\mathfrak{B}$ may be embedded.*

Proof. The proof is carried on by complementing $\mathfrak{B}$ until we obtain $\mathfrak{A}$ by means of MacNeille's cuts [25]. Cf. [37] and [43], p. 152–155.

LEMMA 2. *If $\mathfrak{A}$ is a Boolean σ-algebra, then there exists σ-field of sets $\mathfrak{B}$ and σ-filter $\mathfrak{F}$ in the algebra $\mathfrak{B}$ such that $\mathfrak{A}$ is σ-isomorphic with the quotient algebra $\mathfrak{B}/\mathfrak{F}$.*

Proof. Cf. [43], p. 117.

LEMMA 3. *If $\mathfrak{A}$ is a denumerable Boolean $\Phi\sigma$-algebra with a denumerable family Φ of the marked sequences then there exist a field of sets $\mathfrak{B}$ and a family Φ_1 of the marked sequences, composed of the sets in $\mathfrak{B}$, and there exists a generalized σ-isomorphism of the algebra $\mathfrak{A}$ onto the algebra $\mathfrak{B}$.*

Proof. The detailed proof of this lemma, based on the lemmas 1 and 2, is rather longish. We omit it here. Cf. [37].

2. The Lindenbaum Q-algebra as a Boolean $\Phi\sigma$-algebra and the construction of the model

Assume that $X \subset S$ and $X \in Con$. Like in the method of Rasiowa-Sikorski we shall construct a Lindenbaum Q-algebra $L/_{\sim}$.

LEMMA 4. *The algebra $L/_{\sim}$ is a denumerable Boolean $\Phi\sigma$-algebra with a denumerable family Φ of multiple sequences.*

Proof. We compose the family Φ taking the family of all multiple sequences of the form:

$$\{|A(v_{n_1}//v_{p_1}, \ldots, v_{n_k}//v_{p_k})| : p_1, \ldots, p_k \in \omega\}$$

where $A(v_{n_1}, \ldots, v_{n_k})$ is a formula with the free variables $v_{n_1}, \ldots, v_{n_k}$ and the formula $A(v_{n_1}//v_{p_1}, \ldots, v_{n_k}//v_{p_k})$ is obtained from $A(v_{n_1}, \ldots, v_{n_k})$ as it was described in the previous Chapter VI. We shall demonstrate that Φ satisfies the conditions (i)–(vii) of the definition of the family of marked sequences. The conditions (i)–(v) follow from the definition of the set of all formulas. From lemma 4 in Chapter VI it follows that the conditions (vi) and (vii) are also satisfied. It is obvious now that the family Φ is denumerable.

Q. E. D.

LEMMA 5. *If $|A_0| \neq 0$, then in the Q-algebra $L/_{\sim}$ there exists a $\Phi\sigma$-ultrafilter $\mathfrak{F}$, such that $|A_0| \in \mathfrak{F}$.*

Proof. From lemma 3 we infer that there exists a $\Phi_1\sigma$-field of sets of the subsets Z of a certain set W, and that there exists the function f such that

(1) f is a generalized σ-isomorphism of the algebra $L/_{\sim}$ onto the algebra $\mathfrak{B}$.

Assume that

(2) $|A_0| \neq 0$.

From (1) and (2) it follows that there exists $z \in W$ such that

(3) $z \in f(|A_0|)$.

Suppose that

(4) $\mathfrak{F}_1 = \{Z \in \mathfrak{B} : z \in Z\}$.

Hence and in view of the fact that $\mathfrak{F}_1$ is a filter generated by the unit set, it follows that

(5) $\mathfrak{F}_1$ is a $\Phi_1\sigma$-ultrafilter in $\mathfrak{B}$.

Assume that

(6) $\mathfrak{F}_0 = \{|A| \in L/_\sim\colon$ there exists $Z \in \mathfrak{F}_1$, such that $f(|A|) = Z\}$.

From (1), (2), (5) and (6) we infer that

(7) $\mathfrak{F}_0$ is a $\Phi\sigma$-ultrafilter and $|A_0| \in \mathfrak{F}_0$.

Q. E. D.

The construction of the model is in this case analogous to that in the method of Rasiowa-Sikorski. The only difference is that in place of $\mathfrak{F}$ we put $\mathfrak{F}_0$ which was mentioned in lemma 5. For $\mathfrak{F}_0$ there also holds equivalence (7) stated in the previous Chapter in course of proof of theorem 3; this condition entails that the model constructed is the model of the set X.

VIII. The methods of Reichbach and of Słupecki—Pogorzelski of proving the extended completeness theorem

The methods to be discussed below are modifications of Henkin's construction.

The results of J. Reichbach appeared in print in 1955 in paper [36], but were presented in public as early as 1952, independently of the results of Beth, Rieger, Rasiowa and Sikorski (cf. [36], p. 213). The paper [47] of J. Słupecki and W. A. Pogorzelski was published much later, i.e., in 1961. We want to compare it with the work of J. Reichbach because there plays an essential role the concept of (proper) prime ideal included in the set of all formulas, which was introduced by J. Reichbach. The central problem in this method is to give the proof of the theorem on the existence of prime ideals. Here J. Reichbach gives a purely combinatoric proof of the theorem making up a construction similar to that of G. Hasenjaeger. J. Słupecki and W. A. Pogorzelski, however, utilize for this purpose the so-called rejection function of Łukasiewicz; this concept was introduced by J. Słupecki in his paper [46] in 1959.

Since the formalization of the *LPC* assumed in both of the discussed papers is essentially different from the formalization introduced in Chapter I of the present paper, we are thus obliged to change considerably some of the constructions.

In both papers [36] and [47] was proved the narrower completeness theorem (theorem 1 in the present paper) for the *LPC* without identity; at the same time the authors make use of the concept of prime ideal which, however, is differently defined by each of them. We shall demonstrate that the method may be transferred to the proof of the Gödel-Malcev's completeness theorem, i.e., theorem 3, for the *LPC* with identity.

1. Reichbach's method

Let us symbolize by $L(C)$, like previously, the language obtained from L by adding to it the set $C = \{c_i: i \in \omega\}$ of individual constants.

The set J of the formulas expressed in the language $L(C)$ is the (proper) prime ideal (according to Reichbach [36]) provided for any formulas $A_1, \ldots, A_m, A$ the following conditions are fulfilled:

(i) If $A_1, A_2, \ldots, A_m \in J$, then non $\vdash A_1 \vee A_2 \vee \ldots \vee A_m$.

(ii) If $A \notin J$ then there exist $A_1, A_2, \ldots, A_n \in J$ such that $\vdash A_1 \vee A_2 \vee \ldots \vee A_n \vee A$.

(iii) If $(x)A(x) \in J$, then there exists $k \in \omega$ such that $A(x/c_k) \in J$.

Every prime ideal J determines a model

$$\mathfrak{M}(J, C) = \langle U, \{s_i: i \in \omega\}, P, Q, Id\rangle$$

defined by the conditions:

(i) $U = C/_\sim$,

(ii) $s_i = c_i^\sim$,

(iii) $c_i^\sim \in P$ iff $P(c_i) \notin J$,

(iv) $\langle c_i^\sim, c_j^\sim\rangle \in Q$ iff $Q(c_i, c_j) \notin J$,

(v) $\langle c_i^\sim, c_j^\sim\rangle \in Id$ iff $c_i \doteq c_j \notin J$.

The relation $\sim$ is defined in C by the condition: $c_i \sim c_j$ iff $c_i \doteq c_j \notin J$. With respect to the fact that $\sim$ is an equivalence relation $\mathfrak{M}(J, C)$ is a model with an absolute concept of identity.

It is easy to prove the following:

LEMMA 1. *If J is a prime ideal then for any $A, B \in S$ the following conditions are satisfied*:

(i) $A \in J$ *and* $B \in J$ *iff* $A \vee B \in J$,

(ii) $A \notin J$ *iff* $\neg A \in J$,

(iii) $(x)A(x) \in J$ *iff there exists* $k \in \omega$ *such that* $A(x/c_k) \in J$.

LEMMA 2. *If* $\overline{c^\sim} = \langle c_0^\sim, c_1^\sim, \ldots\rangle$, *then for any* $A \in S$

$$\mathfrak{M}(J, C) \models A[\overline{c^\sim}] \quad \textit{iff} \quad A \notin J.$$

Proof. The proof is to be carried on along the lines analogous to these of the proof of lemma 1 in Henkin's method, i.e., by induction on the construction of A. We base on lemma 1 at the same time.

Q. E. D.

LEMMA 3. *If* $X \subset S$, *then from the fact that the sequence*

(+) $A_1, A_2, \ldots$

has the following properties:

(1) *for any* $n \in N$, *non* $X \vdash A_1 \vee A_2 \vee \ldots \vee A_n$,

(2) *if* B *is not the term of the sequence* (+), *then there exists* $r \in N$ *such that* $\vdash A_1 \vee A_2 \vee \ldots \vee A_r \vee B$,

(3) *if for some* D, $A_n = (x)D(x)$, *then there exists* $k \in \omega$ *such that* $A_{n+1} = D(x/c_k)$,

it follows that the set of terms of the sequence (+) *is a prime ideal.*

Proof. The thesis of lemma 3 is an apparent conclusion from the definition of prime ideal and from the remark that if non $X \vdash A$, then what is more non $\vdash A$.

Q. E. D.

LEMMA 4. *If non* $X \vdash A^+$, *then there exists a prime ideal* J, *such that* $A^+ \in J$.

Proof. Assume that

(1) non $X \vdash A^+$

and, moreover,

(2) $B_1, B_2, \ldots$

is the sequence of all formulas of in the language $L(C)$ and such that $B_1 = A^+$.

We define the sequence of formulas

(3) $D_1, D_2, \ldots$

by means of the conditions:

(3.1) $D_1 = B_1 = A^+$,

(3.2) $D_{n+1} = B(x/c_k)$, if $D_n = (x)B(x)$ and c_k is a constant that does not occur in $D_1, \ldots, D_n$,

$D_{n+1} = B_l$, if D_n does not begin with an universal quantifier and B_l is the first term of the sequence (2), which is not yet included in the sequence (3).

Let the sequence

(4) $A_1, A_2, \ldots$

be a subsequence of the sequence (3), and defined in the following way:

(4.1) $A_1 = A^+$,

(4.2) A_{n+1} is the first term to follow the term A_n in the sequence (3), and such that non $X \vdash A_1 \vee \ldots \vee A_n \vee A_{n+1}$.

We shall prove that the set of terms of the sequence (4) is a prime ideal.

From (1) and (4.2) it follows that

(5) the sequence (4) has the property (1) of lemma 3.

Since in (3) there occur all the formulas, then there exist formulas that are not included in (4). Thus

(6) the sequence (4) has the property (2) defined in lemma 3.

We shall still prove that the sequence (4) has the property (3) defined in lemma 3. Aiming at proving this, let us assume that

(7) $A_n = (x)B(x)$.

From (4) we infer that

(8) non $X \vdash A_1 \vee \ldots \vee A_n$

From the definition of (4) as a subsequence of (3) it follows that

(9) there exists $t \in N$ such that $A_n = D_t$

and such

(10) $D_{n+1} = B(x/c_k)$, where c_k does not occur in $D_1, \ldots, D_m$.

It follows from (8) that

(11) non $X \vdash A_1 \vee \ldots \vee A_{n-1} \vee D_t \vee D_{t+1}$.

Hence and from (9) we deduce that

(12) $A_{n+1} = D_{t+1}$.

From (5), (6), (7) and (12) by virtue of lemma 3 we infer that the set of terms of (4) is a prime ideal, which satisfies the thesis of lemma 4.

Q. E. D.

LEMMA 5. *If* $X \in Con$, *then there exists prime ideal* J *such that* $J \subset S - X$.

Proof. If $X \in Con$, then there exists $A^+ \in S$ such that non $X \vdash A^+$. The ideal J constructed in the proof of lemma 4 has such a property that $J \subset S - X$. Indeed, let $A_0 \in X$ and let us moreover suppose that $A_0 \in J$. From the definition of ideal J it follows that non $X \vdash A_0$ which contradicts the fact that $A_0 \in X$.

Q. E. D.

2. The method of Słupecki-Pogorzelski of proving the theorems on existence of prime ideals

Let us recall the axioms of the consequence theory of A. Tarski [50] which has its primitive terms a set Z and the consequence function $Cn: 2^Z \to 2^Z$. Let T and U be the subsets of Z.

(T1) $\overline{\overline{Z}} \leqslant \aleph_0$,

(T2) $T \subset Cn\, T \subset Z$,

(T3) if $T \subset U$ then $Cn\, T \subset Cn\, U$,

(T4) $Cn\, Cn\, U \subset Cn\, U$,

(T5) $Cn\, U = \bigcup_{U_f} Cn\, U_f$, where $U_f \in Fin\, U$,

(T6) there exists $z \in Z$ such that $Cn\{z\} = Z$.

Observe that the axioms (T1)–(T6) allow us to prove Lindenbaum's theorem (cf. [50]).

The function $Cn'\colon 2^S \to 2^S$ will be called the rejection function of Łukasiewicz provided that for any $X \subset S$ and $A \in S$, $A \in Cn'X$, if and only if there exists a sequence $A_1, \ldots, A_n$ of formulas such that $A_n = A$, and for any $i \leqslant n$ one of the following conditions is satisfied:

(i) $A_i \in X$,

(ii) there exists $j < i$, and there exists $k < i$ such that $A_i = A_j \vee A_k$,

(iii) there exists $j < i$ such that $A_j = (x)B(x)$ and $A_i = B(x/c_k)$, where c_k is the individual constant that does not occur in $A_1, \ldots, A_{i-1}$,

(iv) there exists $j < i$ such that $\vdash A_i \to A_j$.

Lemma 6. *If we put S in place of Z in the axioms* (T1)–(T6), *and we put Cn' in place of Cn, then* (T1)–(T6) *will be true.*

Proof. The justification of (T1)–(T5) is simple in view of the definition of S and of Cn'. (T6) holds in view of the fact that if $\vdash A$, then by virtue of the condition (iv) of the definition of Cn' we obtain $Cn'\{A\} = S$.

Q. E. D.

On the grounds of the remark following Tarski's axioms, and basing on lemma 6 we infer that the analogue of Lindenbaum's theorem for the function Cn' holds, viz.,

Lemma 7. *If $X \in Con'$, then there exists $Y \supset X$ such that $Y \in Syst' \cap Con' \cap Com'$.*

In lemma 7 by symbols *Syst'*, *Con'* and *Com'* we symbolize the family of all systems, consistent and complete sets of formulas with respect to Cn'.

The subsequent lemma establishes some properties of the function Cn'. We shall refer to them in some subsequent proofs.

LEMMA 8.

(a) *If* $A \in Cn'\{B\}$ *then,* $(x_1)\dots(x_n)A \to B$, *where* $x_1, \dots, x_n$ *are all free variables in the formula* A.

(b) *If* $(x_1)\dots(x_m)A \in Cn'X$, *then* $A \in Cn'X$.

(c) *If* $X = \{A_1, \dots, A_m\}$, *then* $Cn'X = Cn'\{A_1 \vee A_2 \vee \dots \vee A_m\}$.

Proof. The easy proof basing on the definition of the function Cn' is here omitted.

LEMMA 9. *If*

(1) $X \in Con$,

(2) $NX = \{A \in S\colon \neg A \in X$, *or there exists* $B \in X$ *such that* B *does not begin with the negation and* $A = \neg B\}$

then

(3) $NX \in Con'$.

Proof. Assume (1) and (2). In case when $X = \emptyset$ we infer that $Cn'NX = Cn'\emptyset = \emptyset$. Then in this case (3) holds. Suppose that $X \neq \emptyset$ and assume indirectly that

(4) $Cn'NX = S$.

From (4) it follows that there exists a sentence A such that

(5) $A \in Cn'NX$ and $\neg A \in Cn'NX$.

From (5), from Cn' being finitistic and from lemma 8(c) we obtain that there exist formulas $A_1, \dots, A_n, B_1, \dots, B_m \in NX$ such that

(6) $A \in Cn'\{A_1 \vee \dots \vee A_n\}$ and $\neg A \in Cn'\{B_1 \vee \dots \vee B_m\}$.

This and lemma 8(a) yield that

(7) $\vdash A \to A_1 \vee \dots \vee A_n$ and $\vdash \neg A \to B_1 \vee \dots \vee B_m$.

Hence we deduce that

(8) $\vdash \neg A_1 \wedge \neg A_2 \wedge \dots \wedge \neg A_n \to \neg A$ and $\vdash \neg B_1 \wedge \dots \wedge \neg B_m \to A$.

From the definition of the formulas $A_1, \dots, A_n, B_1, \dots, B_m$ and in view of (2) and (8) we infer that

(9) $X \vdash A$ and $X \vdash \neg A$.

From (9) it follows that

(10) $X \notin Con$.

From this we infer that the assumption (4) is impossible, then

(11) $NX \in Con'$.

Q. E. D.

We have prepared the sufficient theoretic apparatus to pass now to proving lemma 5 on the existence of prime ideals, utilizing the method of Słupecki-Pogorzelski.

Proof of lemma 5. From the assumption that $X \in Con$ it follows by virtue of lemma 9 that

(1) $NX \in Con'$.

Hence and from lemma 7 we deduce that there exists Y such that

(2) $Y \in Syst' \cap Con' \cap Com'$ and $NX \subset Y$.

We shall demonstrate that

(3) $X \cap Y = \emptyset$.

Assume that there exists A such that

(3.1) $A \in X \cap Y$.

From the definition of the set NX and from (2) it follows in view of (3.1) that

(3.2) $\neg A \in Y$.

From (2), (3.1) and (3.2) and the condition (ii) of the definition of Cn' we infer that

(3.3) $A \vee \neg A \in Y$.

Hence and from the condition (iv) of the definition of Cn' we deduce that

(3.4) $Y \notin Con'$.

The condition (3.4) contradicts the condition (2). We have then proved (3). Now we shall justify that Y is a prime ideal. Assume, aiming at proving this, that

(4) $A_1, \ldots, A_n \in Y$.

Hence and from the condition (ii) of the definition of Cn' it follows that

(5) $A_1 \vee \ldots \vee A_n \in Y$.

This, together with the condition (iv) of the definition of Cn' yields that

(6) non $\vdash A_1 \vee \ldots \vee A_n$.

Hence and from (4) we infer that

(7) Y satisfies the condition (i) of the definition of prime ideal.

Now we shall demonstrate that

(8) if $A \notin Y$ then $\neg A \in Y$.

Assume then, that

(8.1) $A \notin Y$

and suppose indirectly that

(8.2) $\neg A \notin Y$.

From (8.1) in view of the completeness of Y we infer that

(8.3) $Cn'(Y \cup \{A\}) = S$.

Hence it follows that there exists $Z \in Fin\ (Y \cup \{A\})$ such that

(8.4) $(x_1)\ldots(x_n)\neg A \in Cn'Z$

where $x_1, \ldots, x_n$ are free variables in A.

In view of lemma 8(b) it is easy to check that the condition $A \notin Z$ implies contradiction with condition (8.1).

Assume then, that

(8.5) $A \in Z$.

By B we denote the alternative composed of all formulas belonging to the set $Z-\{A\}$. In view of (8.4) and (8.5) by virtue of lemma 8(c), we infer that

(8.6) $(x_1)\ldots(x_n)\neg A \in Cn'\{B \vee A\}$.

Hence and from lemma 8(a) it results that

(8.7) $\vdash (x_1)\ldots(x_n)\neg A \to B \vee A$.

This yields that

(8.8) $\vdash (x_1)\ldots(x_n)\neg A \to B$.

Hence, by definition of Cn' it follows that

(8.9) $(x_1)\ldots(x_n)\neg A \in Cn'\{B\}$.

From the definition of B and from (8.9) it follows by virtue of lemma 8(c), that

(8.10) $(x_1)\ldots(x_n)\neg A \in Cn'(Z-\{A\})$.

This yield that

(8.11) $(x_1)\ldots(x_n)\neg A \in Y$.

Hence and from (2) it follows by lemma 8(b), that

(8.12) $\neg A \in Y$

which contradicts (8.2). The contradiction thus obtained completes the proof of (8).

We obtain from (8) as an immediate conclusion, that

(9) if $B \notin Y$, then there exist $A_1, \ldots, A_m \in Y$ such that $\vdash A_1 \vee \vee \ldots \vee A_n \vee B$.

Assume now, that

(10) $(x)A(x) \in Y$.

In view of (2), (10) and of the definition of Cn' we infer that there exists a constant c such that

(11) $A(x/c) \in Y$.

Lines (3), (7), (9), (10) and (11) yield that

(12) Y is a prime ideal that satisfies the conditions of the thesis of lemma 5.

Q. E. D.

IX. The method of ultraproducts of proving the compactness theorem

The subsequent definition of the reduced ultraproduct of family of models is due to J. Łoś [24]. In the paper [24] there is contained in an implicit way, the fundamental lemma on ultraproducts (i.e., lemma 2 below). The idea of utilizing the ultraproduct construction for proving the compactness theorem is due to A. Tarski in 1958. The proof was published in the paper by A. Morel, D. Scott and A. Tarski [29], cf. also [8].

Let $\{\mathfrak{M}_i: i \in I\}$ be a family of models of the language L with identity, i.e., of models $\mathfrak{M}_i = \langle U_i, P_i, Q_i, Id_i\rangle$.

Let $\mathfrak{F}$ be an ultrafilter in I, and let $\prod_{i\in I} U_i$ be a generalized Cartesian product of the sets U_i. Thus $\prod_{i\in I} U_i = \{g: g: I \to \bigcup_{i\in I} U_i \text{ and } g(i) \in U_i\}$.

LEMMA 1.

(a) *The relation* $\sim$ *defined in* $\prod_{i\in I} U_i$ *by means of the condition:* $f \sim g$ *iff* $\{i \in I: \langle f(i), g(i)\rangle \in Id_i\} \in \mathfrak{F}$ *is an equivalence.*

(b) *The relation* $\sim$ *defined in* (a) *is a congruence relation with respect to every relation* R *defined by means of the condition:*

$$\langle f_1, \ldots, f_n\rangle \in R \text{ iff } \{i \in I: \langle f_1(i), \ldots, f_n(i)\rangle \in R_i\} \in \mathfrak{F}$$

where R_i *is an arbitrary* n*-ary relation in* U_i.

Proof. The simple proof which is based on the properties of ultrafilter and on the definitions introduced, is here omitted. (Cf. [2], p. 87–88.)

From lemma 1 it follows that the below definitions are correct.

$$\prod_{i\in I} U_i/_{\mathfrak{F}} = \{g^{\sim}: g \in \prod_{i\in I} U_i\}$$

where $g^{\sim} = \{f: f \sim g\}$.

The name of the ultraproduct of the family $\{\mathfrak{M}_i: i \in I\}$ will be given to the model $\prod_{i\in I} \mathfrak{M}_i/_{\mathfrak{F}} = \langle U, P, Q, Id\rangle$ where

(i) $U = \prod_{i\in I} U_i/_{\mathfrak{F}}$,

(ii) $f^{\sim} \in P$, iff $\{i \in I: f(i) \in P_i\} \in \mathfrak{F}$,

(iii) $\langle f^{\sim}, g^{\sim}\rangle \in Q$, iff $\{i \in I: \langle f(i), g(i)\rangle \in Q_i\} \in \mathfrak{F}$,

(iv) $\langle f^{\sim}, g^{\sim}\rangle \in Id$, iff $\{i \in I: \langle f(i), g(i)\rangle \in Id_i\} \in \mathfrak{F}$.

Let us accept the following notations:

$$[\overline{u}] = \langle f_1^{\sim}, f_2^{\sim}, \ldots\rangle$$

$$[\overline{u(i)}] = \langle f_1(i), f_2(i), \ldots\rangle$$

$$[\overline{u(n/g)}] = \langle f_1^{\sim}, f_2^{\sim}, \ldots, f_{n-1}^{\sim}, g^{\sim}, f_{n+1}^{\sim}, \ldots\rangle .$$

With these notations accepted the fundamental lemma on ultraproducts holds.

LEMMA 2 (Łoś [24]). *For an arbitrary formula* A

$$\prod_{i\in I} \mathfrak{M}_i/_{\mathfrak{F}} \models A[\overline{u}], \quad \textit{iff} \quad \{i \in I: \mathfrak{M}_i \models A[\overline{u(i)}]\} \in \mathfrak{F}.$$

Proof. I. The lemma holds for atomic formulas in view of the conditions (i)–(iv) of the definition of an ultraproduct.

II. Assume that

(1) the lemma holds for the formulas B and D.

Accept the following notations:

(2.1) $F_B = \{i \in I: \mathfrak{M}_i \models B[f_1(i), f_2(i), \ldots]\}$

(2.2) $F_D = \{i \in I: \mathfrak{M}_i \models D[f_1(i), f_2(i), \ldots]\}$.

From the definition of satisfaction it follows that the condition

(3) $\prod_{i\in I} \mathfrak{M}_i/_{\mathfrak{F}} \models B \wedge D[\overline{u}]$

is equivalent to the conjunction of the below conditions:

(4.1) $\prod_{i\in I} \mathfrak{M}_i/_{\mathfrak{F}} \models B[\overline{u}]$

(4.2) $\prod_{i\in I} \mathfrak{M}_i/_{\mathfrak{F}} \models D[\overline{u}]$.

Assumption (1) together with (4.1) and (4.2) yield that (4.1) is equivalent to the fact that

(5.1) $F_B \in \mathfrak{F}$

and (4.2) is equivalent to the fact that

(5.2) $F_D \in \mathfrak{F}$.

From the definition of the ultrafilter it follows that (5.1) and (5.2) is equivalent to the fact that

(6) $F_B \cap F_D \in \mathfrak{F}$.

From the definition of satisfaction and from (2) it follows that (6) is equivalent to the condition:

(7) $\{i \in I\colon \mathfrak{M}_i \models B \wedge D[\overline{u(i)}]\} \in F$.

From (3) and (7) we deduce that

(8) the lemma holds for any formula of the form $B \wedge D$.

Assume now that

(9) $\prod_{i \in I} \mathfrak{M}_i/_{\mathfrak{F}} \models \neg B[\overline{u}]$.

From the definition of satisfaction and from the assumption (1) it follows that (9) is equivalent to the fact that

(10) $F_B \notin \mathfrak{F}$.

From the definition of the ultrafilter we infer that (10) is equivalent to the fact that

(11) $I - F_B \in \mathfrak{F}$.

From the definition of the set F_B we infer that (11) is equivalent to the fact that

(12) $\{i \in I\colon \mathfrak{M}_i \models B[\overline{u(i)}]\} \in F$.

We obtain from (9) and (12) that

(13) the lemma holds for the formulas of the form $\neg B$.

Assume now that

(14) $A = (Ev_n)B$

and accept the notation

(15) $F = \{i \in I\colon \mathfrak{M}_i \models (Ev_n)B[\overline{u(i)}]\}$

Assume moreover, that

(16) $\prod_{i \in I} \mathfrak{M}_i/_{\mathfrak{F}} \models (Ev_n)B[\overline{u}]$.

From (16) and from the definition of satisfaction we infer that

(17) there exists $g \in \prod_{i \in I} U_i$ such that $\prod_{i \in I} \mathfrak{M}_i/_{\mathfrak{F}} \models B[\overline{u(n/g)}]$.

Assume that

(18) $F_1 = \{i \in I\colon \mathfrak{M}_i \models B[\overline{u(n/g)(i)}]$.

From (18) and from the inductive assumption (1) we infer that

(19) $F_1 \in \mathfrak{F}$.

From the accepted notations it follows that

(20) $\overline{u(n/g)(i)} = \overline{u(i)(n/g(i))}$.

We deduce from (15), (18) and (20) that

(21) $F_1 \subset F$.

From (19) and (21) it follows that

(22) $F \in \mathfrak{F}$.

Assume now (22). It follows hence that

(23) if $i \in F$, then $\mathfrak{M}_i \models (Ev_n)B[\overline{u(i)}]$.

From (23) there follows the existence of $g_i \in U_i$, which is such that

(24) $\mathfrak{M}_i \models B[\overline{u(i)(n/g_i)}]$.

Let us construe now the function $g_0 \in \prod_{i \in I} U_i$, such that

(25) $$g_0(i) = \begin{cases} g_i, & \text{for } i \in F \\ \text{arbitrary}, & \text{for } i \notin F. \end{cases}$$

From (15) and (25) we obtain that

(26) $F \subset \{i \in I: \mathfrak{M}_i \models B[\overline{u(n/g_0)(i)}]\} = F_3$.

We deduce from (22) and (26) that

(27) $F_3 \in \mathfrak{F}$.

It follows from (27) and from the inductive assumption that

(28) $\prod_{i \in I} \mathfrak{M}_i/_{\mathfrak{F}} \models B[\overline{u(n/g_0)}]$.

We infer from (28) and from the definition of satisfaction that

(29) $\prod_{i \in I} \mathfrak{M}_i/_{\mathfrak{F}} \models (Ev_n)B[\overline{u}]$.

From (15), (16), (22) and (28) it follows that

(30) the lemma holds for the formulas of the form $(Ev_n)B$.

Q. E. D.

LEMMA 3 (Tarski [49]). *An arbitrary family $\mathfrak{R} \subset 2^I$ of the subsets of non-empty set I may be extended to an ultrafilter on I, if and only if $\mathfrak{R}$ has a finite intersection property, i.e., for an arbitrary $n \in N$, if $Z_1, Z_2, \ldots, Z_n \in \mathfrak{R}$, then $Z_1 \cap Z_2 \cap \ldots \cap Z_n \neq \emptyset$.*

Proof. Lemma 3 is a particular case of the analogous theorem for arbitrary Boolean algebras (and not only for the fields of sets). Regarding this generalization and its proof cf., e.g., [2], p. 16.

We shall prove now compactness theorem 5. Assume that

(1) $X \in S$.

Let us accept the following notations:

(2) $I = Fin X$,

(3) $\hat{i} = \{i' \in I: i \subset i'\}$,

(4) $\mathfrak{R} = \{\hat{i}: i \in Fin X\}$.

Assume that

(5) for any $i \in I$ there exists a model $\mathfrak{M}_i$ of the set i.

From (3) it follows that for every $n \in N$

(6) $i_1 \cup i_2 \cup \ldots \cup i_n \in \hat{i}_1 \cap \hat{i}_2 \cap \ldots \cap \hat{i}_n$.

We deduce from (6) that

(7) $\hat{i}_1 \cap \hat{i}_2 \cap \ldots \cap \hat{i}_n \neq \emptyset$.

From (4), (7) and from lemma 3 we infer that

(8) there exists an ultrafilter $\mathfrak{F}$ on I, which is an extension of $\mathfrak{R}$.

Let

(9) $\prod_{i \in I} \mathfrak{M}_i/_{\mathfrak{F}}$

be an ultraproduct of the family $\{\mathfrak{M}_i : i \in I\}$ defined in (5).

Assume that

(10) $A \in X$.

It follows therefrom that

(11) $\{A\} \in I$.

From (11) in turn, we obtain in view of (3), that

(12) $\mathfrak{M}_{\{A\}} \models A$.

From (11) it follows that for any $i_0 \supset \{A\}$

(13) $\mathfrak{M}_{i_0} \models A$.

From the definition of the operation $\hat{}$ in (3), from (13) and from the definition of satisfaction we obtain that

(14) $\{\hat{A}\} = \{i' \in I : \{A\} \subset i'\} \subset \{i \in I : \mathfrak{M}_i \models A\}$.

We infer from (8) and from (14) that

(15) $\{i \in I : \mathfrak{M}_i \models A\} \in \mathfrak{F}$.

From (15) and from lemma 2 we deduce that

(16) $\prod_{i \in I} \mathfrak{M}_i/_{\mathfrak{F}} \models A$.

It follows from (10) and (16) that

(17) $\prod_{i \in I} \mathfrak{M}_i/_{\mathfrak{F}} \models X$

which completes the proof of theorem 5.

X. Comparison of the discussed methods

1. Connections between Gödel's original method and the method of ultraproducts

The considerations contained in this section are based on the ideas of J. N. Crossley as presented by J. L. Bell and A. B. Slomson in [2], p. 233–236.

First we shall introduce some concepts related to ultraproducts.

We say that the filter $\mathfrak{F}$ on set I is principal if and only if there exists $I_f \in Fin I$ such that $I_f \in \mathfrak{F}$. Otherwise the filter is called non-principal.

LEMMA 1. *If I is a non-finite set, then there exists a non-principal ultrafilter on I.*

Proof. The family of all complements of the finite subsets of the set I has the finite intersection property. According to lemma 3 of Chapter IX the family is capable of being extended to an ultrafilter which satisfies the thesis of lemma 1.

Q. E. D.

Let $\mathfrak{F}$ be an ultrafilter on N. Let us consider the power ω^N. A function $*$ is defined on ω and performable in ω^N, and such that for i

$$^*(i) = i^*$$

and also for any $n \in N$

$$i^*(n) = i .$$

Let $d\colon \omega \to \omega^N/\mathfrak{F}$ be the canonical embedding of ω in $\omega^N/\mathfrak{F}$, i.e., the function defined by the following:

$$d(i) = i^{*\sim}$$

where $i^{*\sim} = \{f \in \omega^N\colon i^* \sim_{\mathfrak{F}} f\}$.

The set of values of the function d may be identified with the set of sequences $\{d(i)\colon i \in \omega\}$ where

$$d(i) = \langle i, i, i, \ldots\rangle^{\sim} .$$

Assume that

$$d(\omega) = \langle d(0), d(1), d(2), \ldots\rangle \in (\omega^N/\mathfrak{F})^{\omega}$$

and that

$$d[\omega] = \{d(0), d(1), d(2), \ldots\} ,$$

i.e., that $d[\omega]$ is the set of terms of the sequence $d(\omega)$.

Presenting the connections between Gödel's method for proving the narrower completeness theorem with the method of ultraproducts, we shall refer to the paper [53], and particularly to lemmas 8 and 9 as well as to the notations there introduced. We shall demonstrate another proof of the crucial lemma 9 according to which the sentence $(\alpha)(E\beta)A(\alpha, \beta)$ is either refutable, or there exists a model verifying it. Proving lemma 9 we shall distinguish, like previously, between the following two cases:

1° There exists $n \in N$ such that Z_n is refutable in the classical propositional calculus. In this case, reasoning like in [53], we infer that $(\alpha)(E\beta)A(\alpha,\beta)$ is the refutable formula.
2° For every $n \in N$, Z_n is satisfiable. Hence it follows by virtue of lemma 11 of Chapter I, that for an arbitrary $n \in N$ there exists a model $\mathfrak{M}_n$ of the language L in the domain ω, such that

(1) $\mathfrak{M}_n \models A_n(x_0, \ldots, x_{n\cdot t})[0, 1, 2, \ldots]$.

From the definition of the formula A_n considered as a certain conjunction, it follows that for $j > k$

(2) $\vdash A_j \to A_k$.

From (1) and (2) it follows that for $j > k$

(3) $\mathfrak{M}_j \models A_k(x_0, \ldots, x_{k\cdot t})[0, 1, 2, \ldots]$.

From (3) we deduce that

(4) $F = \{j \in N\colon \mathfrak{M}_j \models A_k(x_0, \ldots, x_{k\cdot t})[0, 1, 2, \ldots]\} \in Fin\, N$.

From (4) and from lemma 1 it follows that

(5) there exists a non-principal ultrafilter $\mathfrak{F}$ on N such that $F \notin \mathfrak{F}$.

Assume that

(6) $\mathfrak{N} = \prod_{i \in N} \mathfrak{M}_i/_{\mathfrak{F}}$

is a ultraproduct of the family of models defined in (1).

From (4), (5), (6) and from lemma on ultraproduct of Łoś we obtain that for every $n \in N$

(7) $\mathfrak{N} \models A_n(x_0, \ldots, x_{n\cdot t})[d(\omega)]$.

Assume that

(8) $\mathfrak{N}_0 = \mathfrak{N} \restriction d[\omega]$

is a model that is obtained from $\mathfrak{N}$ by restricting its domain to the set $d[\omega]$. Observe in view of the definitions quoted above, that

(9) $d(\omega) \in (d[\omega])^\omega$.

From (7), (8) and (9) we infer that

(10) $\mathfrak{N}_0 \models A_n(x_0, \ldots, x_{n\cdot t})[d(\omega)]$.

From (10) and from the definition of A_n we deduce that for an arbitrary $n \in N$

(11) $\mathfrak{N}_0 \models A(\alpha_n, \beta_n)[d(\omega)]$.

Hence, in view of the definition of the sequences α_n, β_n we infer that

(12) $\mathfrak{N}_0 \models (\alpha)(E\beta)A$.

Let us still observe that in [5] there are underlying the connections between Gödel's method and the original method of proof of Herbrand's theorem [16].

2. The connections between Henkin's method and the method of ultraproducts (cf. [20])

Let $\{\mathfrak{M}_i\colon i \in I\}$ be the family of models of the language L with identity and let $C = \prod_{i \in I} U_i$ be the Cartesian product of their domains. Let μ be an ordinal number, large enough that the function

$$c\colon \mu \to C$$

be the mapping „onto". Let also $c(a) = c_a$ for $a < \mu$. Thus

$$c_a\colon I \to \bigcup_{i \in I} U_i$$

and $c(i) \in U_i$.

For any $i \in I$, let $s^{(i)}$ be such a function that

$$s^{(i)}\colon \mu \to U_i$$

defined by the following: $s^{(i)}(a) = c_a(i)$.

Let us enrich the language L with a sequence of arbitrary individual constants of the length μ. This sequence may be identified with the sequence $\{c_a\colon a < \mu\}$ whose set of terms is identical with C.

We shall be concerned with the extension $L(C)$ of the language L. Every model $\mathfrak{M}_i$ of the language L will be now extended so as to include the interpretations of the constants c_a assuming that the element $s^{(i)}(a)$ is the interpretation of the constant c_a in the model $\mathfrak{M}_i$. Thus the i-th model of the language $L(C)$ will have the form

$$\mathfrak{M}_i' = \langle U_i, \{c_a(i)\colon a < \mu\}, P_i, Q_i, Id_i \rangle .$$

The family $\mathfrak{M}_i'$ will be fixed for the time of further considerations.

LEMMA 2. *If*

(1) X^+ *is a set of sentences in the language* $L(C)$,

(2) $X^+ = \{A\colon$ *for any* $i \in I$, $\mathfrak{M}_i \models A\}$,

(3) $X^+ \subset Y$,

(4) $Y \in Syst \cap Con \cap Com$,

then

(5) Y *has the property* $(\mathfrak{H})$ *in the set* C.

Proof. Assume (1)–(4). Let $(Ex)A(x)$ be a sentence in $L(C)$ such that

(6) $(Ex)A(x) \in Y$.

Obviously, $(Ex)A(x)$ is valid in $\mathfrak{M}'_i$ or it is not. Assume then first that for any $i \in I$

(7) $\mathfrak{M}'_i \models (Ex)A(x)$.

From (7) and definition of satisfaction it follows that there exists $u_i \in U_i$ such that

(8) $\mathfrak{M}'_i \models A(x)[u_i]$.

It follows from (8) that

(9) the function u, such that $u(i) = u_i$ is an element of C.

It follows from (9) that there exists $\alpha < \mu$, such that

(10) $u = c_\alpha$.

It is obtained from (7) and (9) that

(11) $\mathfrak{M}'_i \models (Ex)A(x) \to A(x/c_\alpha)$.

From (2) and (11) it is deduced that

(12) $(Ex)A(x) \to A(x/c_\alpha) \in X^+$.

We infer from (3) and (12) that

(13) $(Ex)A(x) \to A(x/c_\alpha) \in Y$.

From (4), (6) and (13) we deduce that

(14) Y has the property $(\mathfrak{H})$ in the set C.

The case when $(Ex)a(x)$ is false in $\mathfrak{M}'_i$ is obvious in view of the thesis $\neg(Ex)A(x) \to ((Ex)A(x) \to A(x/c_\alpha))$.

Q. E. D.

We shall give now, basing on lemma 2, the proof of the compactness theorem which is a certain modification of the proof given in Chapter IX.

Proof. Assume that $\emptyset \neq X \subset S_1$ and suppose that

(1) for every $i \in I = Fin X$ there exists a model $\mathfrak{M}_i$ of the set i.

The assumptions (1) and (2) of lemma 2 yield together with (1), that

(2) $\mathfrak{M}'_i \models i \cup X^+$.

We infer hence and from (1) that

(3) $X \cup X^+ \in Con$.

Hence it follows by virtue of Lindenbaum's theorem that there exists a set Y of sentences in the language $L(C)$ such that

(4) $X \cup X^+ \subset Y$ and $Y \in Syst \cap Con \cap Com$.

Therefrom, in view of lemma 2, we infer that

(5) Y has the property $(\mathfrak{H})$ in the set C.

This and lemma 1 of Chapter V yield that

(6) $\mathfrak{M}(Y, C) \models X$

which completes the proof of theorem 5.

LEMMA 3. *If*

(1) *$\mathfrak{F}$ is an ultrafilter on I while I is the set of indices of the distinguished family $\{\mathfrak{M}_i\}$,*

(2) $S(\mathfrak{F}) = \{A \in S_1 \colon \{i \in I \colon \mathfrak{M}'_i \models A\} \in \mathfrak{F}\}$,

then

(3) *$S(\mathfrak{F}) \in Syst \cap Con \cap Com$ and $S(\mathfrak{F})$ has the property $(\mathfrak{H})$ in the set C.*

Proof. Assume (1) and (2). Suppose now that

(4) $A \in S(\mathfrak{F})$ and $A \to B \in S(\mathfrak{F})$.

From this and from (2) we infer that

(5) $\{i \in I \colon \mathfrak{M}'_i \models A\} \in \mathfrak{F}$

and also that

(6) $\{i \in I \colon \mathfrak{M}'_i \models \neg A \vee B\} \in \mathfrak{F}$.

Since from the definition of satisfaction it follows that

(7) $\{i \in I \colon \mathfrak{M}'_i \models \neg A \vee B\} = \{i \in I \colon \mathfrak{M}'_i \models \neg A\} \cup \{i \in I \colon \mathfrak{M}'_i \models B\}$,

then we deduce basing on (1), (6) and (7), that

(8) $\{i \in I \colon \mathfrak{M}'_i \models B\} \in \mathfrak{F}$.

Hence and from (2) it follows that

(9) $B \in S(\mathfrak{H})$.

This, together with (4) yields that

(10) $S(\mathfrak{H}) \in Syst$.

Along the analogous lines we prove that

(11) $S(\mathfrak{H}) \in Con \cap Com$.

From the fact that $I \in \mathfrak{F}$ and from (2) of lemma 2 we infer that

(12) $X^+ \subset S(\mathfrak{H})$.

Lemma 2 together with (10), (11) and (12) yield the thesis (3) of lemma 3.

Q. E. D.

LEMMA 4. *If the assumptions (1)–(4) of lemma 2 are satisfied, then there exists an ultrafilter $\mathfrak{F}$ on I, such that $Y = S(\mathfrak{F})$, where $S(\mathfrak{F})$ is defined like in lemma 3.*

Proof. Assume (1)–(4). Therefrom it follows that

(5) Y has the property $(\mathfrak{H})$ in the set C.

Let $\mathfrak{F}$ be an ultrafilter on I, such that

(6) $\{i \in I: \mathfrak{M}'_i \models A\} \in \mathfrak{F}$, for any $A \in Y$.

This and lemma 3 yield that

(7) $Y \subset S(\mathfrak{F})$.

From (4) we infer that

(8) $S(\mathfrak{F}) \subset Y$.

Hence and from (7) we obtain that

(9) $S(\mathfrak{F}) = Y$.

Q. E. D.

LEMMA 5. *The model $\mathfrak{M}(S(\mathfrak{F}), C)$ defined in Henkin's lemma 1 of Chapter V is isomorphic with the ultraproduct* $\prod_{i \in I} \mathfrak{M}'_i/_{\mathfrak{F}}$.

Proof. Observe that according to the accepted definitions

(1) $\mathfrak{M}(S(\mathfrak{F}), C) = \langle U, \{s_a: a < \mu\}, P, Q, Id\rangle$

where

(i) $U = C/_{\sim} = \prod_{i \in I} U_i/_{\sim}$

(ii) $s_a = c_a^{\sim}$

(iii) $c_a^{\sim} \in P$ iff $P(c_a) \in S(\mathfrak{F})$

(iv) $\langle c_a^{\sim}, c_\beta^{\sim}\rangle \in Q$ iff $Q(c_a, c_\beta) \in S(\mathfrak{F})$,

(v) $\langle c_a^{\sim}, c_\beta^{\sim}\rangle \in Id$ iff $(c_a \doteq c_\beta) \in S(\mathfrak{F})$

and that the relation $\sim$ is defined in C by the condition: $c_a \sim c_\beta$ iff $(c_a \doteq c_\beta) \in S(\mathfrak{F})$.

Observe also that

(2) $\prod_{i \in I} \mathfrak{M}'_i/_{\mathfrak{F}} = \langle \overline{U}, \{\bar{s}_a: a < \mu\}, \overline{P}, \overline{Q}, \overline{Id}\rangle$

where

(i) $U = C/_{\approx} = \prod_{i \in I} U_i/_{\approx}$

(ii) $\bar{s}_a = c_a^{\approx}$

(iii) $c_a^{\approx} \in \overline{P}$ iff $\{i \in I: c_a(i) \in P_i\} \in \mathfrak{F}$,

(iv) $\langle c_a^{\approx}, c_\beta^{\approx}\rangle \in \overline{Q}$ iff $\{i \in I: \langle c_a(i), c_\beta(i)\rangle \in Q_i\} \in \mathfrak{F}$,

(v) $\langle c_a^{\approx}, c_\beta^{\approx}\rangle \in \overline{Id}$ iff $\{i \in I: \langle c_a(i), c_\beta(i)\rangle \in Id_i\} \in \mathfrak{F}$

and that the relation $\approx$ is defined in C by the condition: $c_a \approx c_\beta$ iff $\{i \in I: \langle c_a(i), c_\beta(i)\rangle \in Id_i\} \in \mathfrak{F}$.

From (1) and from the definition of $S(\mathfrak{F})$ we infer that

(3) $c_a \sim c_\beta$ iff $\{i \in I: \mathfrak{M}'_i \models c_a \doteq c_\beta\} \in \mathfrak{F}$.

Hence and from the definition of interpretation of constants in model $\mathfrak{M}'_i$ it follows that

(4) $c_\alpha \sim c_\beta$ iff $\{i \in I: \langle c_\alpha(i), c_\beta(i)\rangle \in Id_i\} \in \mathfrak{F}$.

This and (2) yield that

(5) $c_\alpha \sim c_\beta$ iff $c_\alpha \approx c_\beta$.

Observe also that

(6) $c_\alpha^\sim \in P$ iff $c_\alpha^\approx \in \overline{P}$.

The analogous property as defined in (6) is valid also for the relations Q and $\overline{Q}$ as well as for Id and $\overline{Id}$.

From the above considerations, we infer that the function f is such that $f(c_\alpha^\sim) = c_\alpha^\approx$ is the isomorphism which was looked for.

Q. E. D.

3. Connection between A. Robinson's method and the method of ultraproducts

Conforming with the previous announcement we shall give now the definition of a ultraproduct of models, utilizing for this purpose corollary 2 of Chapter IV.

Let $\{\mathfrak{M}_i: i \in I\}$ be the family of models of the language L with identity. Let then $\mathfrak{F}$ be an ultrafilter on I and $U = \prod_{i\in I} U_i/_{\mathfrak{F}}$. Let us assign a constant $c_{f^\sim}$ to every $f^\sim \in U$.

Consider now the set of atomic formulas

$$At' = \{P(c_{f^\sim}): f^\sim \in U\} \cup \{Q(c_{f^\sim}, c_{g^\sim}): f^\sim, g^\sim \in U\} \cup \{c_{f^\sim} \doteq c_{g^\sim}: f^\sim, g^\sim \in U\}.$$

We define the set $W = \{w_i: i \in I\}$ of valuations of the set At' in such a way that

$$w_i P(c_{f^\sim}) = 1 \quad \text{iff} \quad f(i) \in P_i$$

$$w_i Q(c_{f^\sim}, c_{g^\sim}) = 1 \quad \text{iff} \quad \langle f(i), g(i)\rangle \in Q_i$$

$$w_i(c_{f^\sim} \doteq c_{g^\sim}) = 1 \quad \text{iff} \quad \langle f(i), g(i)\rangle \in Id_i.$$

From corollary 2 of Chapter IV it follows that there exists a total valuation w of the set At' determined in the unique way by the set W and by the ultrafilter $\mathfrak{F}$. The reduced ultraproduct $\prod_{i\in I} \mathfrak{M}_i/_{\mathfrak{F}}$ is defined now as a model $\langle U, P, Q, Id\rangle$ defined by the conditions:

(i) $U = \prod_{i\in I} U_i/_{\mathfrak{F}}$,

(ii) $f^\sim \in P$ iff $wP(c_{f^\sim}) = 1$

(iii) $\langle f^\sim, g^\sim\rangle \in Q$ iff $wQ(c_{f^\sim}, c_{g^\sim}) = 1$

(iv) $\langle f^\sim, g^\sim\rangle \in Id$ iff $w(c_f \doteq c_g) = 1$.

In view of the way that the valuation w is determined by W and $\mathfrak{F}$, it is easily seen that the above definition is an equivalent paraphrase of the standard definition given in Chapter IX.

4. Connections between the other methods

The lemma below states the interconnection between the method of Henkin and the Reichbach's method.

LEMMA 6. *The set* $Y \in Syst \cap Con \cap Com$ *and* $Y \subset S_1$ *has the property* $(\mathfrak{H})$ *in the set of individual constants* C *if and only if the set* $S_1 - Y$ *is the prime ideal in the set* C *in the sense of Reichbach.*

Proof. I. Assume that $Y \subset S_1$ and

(1) $Y \in Syst \cap Con \cap Com$

and that

(2) Y satisfies the condition $(\mathfrak{H})$ in the set C.

Assume that

(3) $A \vee B \in S_1 - Y$.

Hence and from (1) we infer that (3) is equivalent to the fact that

(4) $\neg A \in Y$ and $\neg B \in Y$.

From (3) and (4) it follows that

(5) $A \vee B \in S_1 - Y$ iff $A \in S_1 - Y$ and $B \in S_1 - Y$.

From (1) we directly obtain that

(6) $\neg A \in S_1 - Y$ iff $A \notin S_1 - Y$.

Assume now that

(7) $(x) A(x) \in S_1 - Y$.

This together with (1), yields that (7) is equivalent to the fact that

(8) $(Ex) \neg A(x) \in Y$.

Hence and from (2) it follows that (8) is equivalent to existence $c \in C$ such that

(9) $\neg A(c) \in Y$.

Therefrom and from (6) we infer that (9) is equivalent to the fact that there exists $c \in C$, such that

(10) $A(c) \in S_1 - Y$.

From (10) and from (7) it follows that

(11) $(x) A(x) \in S_1 - Y$ iff there exists $c \in C$, such that $A(c) \in S_1 - Y$.

From (5), (6) and (11) we deduce that

(12) $S_1 - Y$ is the prime ideal in the sense of Reichbach in the set C.

II. Assuming (12) it is easy now to prove by means of the reasoning analogous to that above, that $Y \in Syst \cap Con \cap Com$ and Y fulfils the condition ($\mathfrak{H}$) in the set C.

Q. E. D.

The two lemmas below explain the relationship between Henkin's method and the method of Beth.

LEMMA 7. *If w is a normal valuation, and $X = \{A \in S_1 : wA = 1\}$, then $X \in Syst \cap Con \cap Com$, and moreover X satisfies the condition ($\mathfrak{H}$) in the set C.*

Proof. It is obvious that

(1) $X \in Syst \cap Con \cap Com$.

Assume now that

(2) $(Ex) A(x) \in X$.

From (1) and (2) it follows that

(3) $(x) \neg A(x) \in S_1 - X$.

Hence and from the definition of the set X we obtain that

(4) $w((x) \neg A(x)) = 0$.

Hence and from the definition of the normal valuation we infer that

(5) there exists $c \in C$ such that $w \neg A(c) = 0$.

Hence we deduce that

(6) $wA(c) = 1$.

This in turn, yields that

(7) $A(c) \in X$.

From this and from (2) it follows that

(8) X has the property ($\mathfrak{H}$) in the set C.

Q. E. D.

LEMMA 8. *If $X \in Syst \cap Con \cap Com$ and $X \subset S_1$ and X has the property ($\mathfrak{H}$) in the set C, then the valuation w defined by the following:*

$$wA = \begin{cases} 1, & \text{if } A \in X \\ 0, & \text{if } A \notin X \end{cases}$$

is a normal valuation.

Proof. The proof of this lemma is similar to that of lemma 7 and therefore we omit it.

LEMMA 9. *In the Lindenbaum $\Phi\sigma$-algebra, $\Phi\sigma$-ultrafilter construed in lemma 5 of Chapter VII is the ultrafilter which preserves the sums of the form:*

$$(+) \qquad \bigcup_{p\in\omega} |A(v_k//v_p)| = |(Ev_k)A(v_k)| .$$

Proof. From the definition of the $\Phi\sigma$-ultrafilter it follows that every $\Phi\sigma$-ultrafilter is a proper prime filter. Hence it follows that every $\Phi\sigma$-ultrafilter is an ultrafilter. In view of the condition (iii) of the definition of the $\Phi\sigma$-ultrafilter, and also by virtue of the way of defining the family Φ it follows that $\mathfrak{F}_0$ preserves the sums $(+)$.

Q. E. D.

The concept of $\Phi\sigma$-ultrafilter is essentially stronger than the concept of the ultrafilter which preserves the sums of the form $(+)$. As it is easy to see from lemma 9, this strengthening is not utilized in the case of proving the completeness theorem by means of Rieger's method.

5. Some final remarks

In the completeness proof for the non-axiomatic version of *LPC* the concept of the Hintikka's set plays an essential role.

The set $H \subset S$ is called the Hintikka's set (cf. [19]) in the set C of individual constants provided that for any $A, B \in S$ the following conditions are satisfied:

(i) if $A \in H$, then $\neg A \notin H$,

(ii) if $A \wedge B \in H$, then $A \in H$ and $B \in H$,

(iii) if $A \vee B \in H$ then $A \in H$ or $B \in H$,

(iv) if $A \rightarrow B \in H$, then $\neg A \in H$ or $B \in H$,

(v) if $A \leftrightarrow B \in H$, then $A \in H$ iff $B \in H$,

(vi) if $(Ex)A(x) \in H$, then $A(c)$, for some $c \in C$,

(vii) if $(x)A(x) \in H$, then $A(c)$, for any $c \in C$.

It is easy task to prove the lemma stating that any Hintikka's set has a model in the domain C. Thus, in the case of these methods of proving the completeness theorem, which make use of the Hintikka's sets, the principal difficulty would be to construct such sets. An unusually charming method of construing the Hintikka's sets and based on the theory of dyadic trees has been developed by Smullyan [48]. This is an elegant simplification of Beth's tableaux and a synthesis of the ideas of Hintikka [19] and Anderson-Belnap [1].

We shall now give a certain simple lemma which will clear up the relationship of Henkin's method and that of Hintikka. First, however, we should accept two definitions.

We shall say that B is an immediate special subformula of A, provided that the following conditions are satisfied:

(i) if $A = A_1 \wedge A_2$ or $A = A_1 \vee A_2$ or $A = A_1 \leftrightarrow A_2$, then $B = A_1$ or $B = A_2$,

(ii) if $A = A_1 \rightarrow A_2$, then $B = \neg A_1$ or $B = A_2$,

(iii) if $A = (x) A(x)$ or $A = (Ex) A(x)$, then $B = A_1(c)$, for some $c \in C$.

We shall say that B is a special subformula of A provided that there exists a sequence of formulas $B_1, \dots, B_n$ such for any $i \leqslant n$ there exists $j < i$ such that B_i is an immediate special subformula of B_j and $B_1 = A$.

LEMMA 10. *If*

(1) *X' is the set of all special subformulas of the elements of a certain set X,*

(2) *Y is the maximal consistent subset of X' with respect to inclusion,*

(3) *Y satisfies the condition ($\mathfrak{H}$) in the set C,*

then

(4) *Y is the Hintikka's set in the set C.*

Proof. The simple proof is omitted here (cf. [48], p. 96).

Assume now that $X \in Con$ and $X \subset S_{(0)}$. Let $A_1, A_2, \dots$ be the sequence of all special subformulas of the elements belonging to X. Let us put that

$X_0 = X$,

$X_{n+1} = X_n \cup \{A_n\}$, if $X_n \cup \{A_n\} \in Con$

$X_{n+1} = X_n$, otherwise.

Let $Y = \bigcup_{n \in \omega} X_n$. The set Y satisfies the assumptions of lemma 10 and, by virtue of this lemma, it is the Hintikka's set. Hence it follows that Y has a model.

Observe that Y has been constructed according to the same rule which was explained in § 3 of Chapter V.

References

[1] ANDERSON A. R., BELNAP N. D.: *A simple proof of Gödel's completeness theorem.* Journal of Symbolic Logic, 24(1959), 320.

[2] BELL J. L., SLOMSON A. B.: *Models and ultraproducts. An introduction.* Amsterdam 1969.

[3] BETH E. W.: *A topological proof of the theorem of Löwenheim-Skolem-Gödel.* Koninklijke Nederlandse Akademie van Wetenschappen. Proceedings of the Section of Sciences, series A, 54(1951), 436–444.

[4] ČECH F.: *On bicompact spaces.* Annales of Mathematics, 38(1937), 823–844.

[5] DREBEN B.: *On the completeness of quantification theory.* Proceedings of the National Academy of Sciences of the USA, 38(1952), 1047–1052.

[6] ENGELKING R.: *Outline of general topology.* Amsterdam 1968.

[7] FEFERMAN S.: *Review of the paper: H. Rasiowa and R. Sikorski, „A proof of the completeness theorem of Gödel"* (Fundamenta Mathematicae, 37(1950), 193–200). Journal of Symbolic Logic, 19(1952), 72.

[8] FRAYNE T., MOREL A. C., SCOTT D.: *Reduced direct products.* Fundamenta Mathematicae, 51(1962/63), 195–228.

[9] GÖDEL K.: *Die Vollständigkeit der Axiome der logischen Funktionenkalküls.* Monatshefte für Mathematik und Physik, 37(1930), 349–360.

[10] GRZEGORCZYK A.: *Zarys logiki matematycznej.* Wydanie drugie. Warszawa 1969.

[11] HALMOS P. R.: *Algebraic logic.* New York 1962.

[12] HASENJAEGER G.: *Eine Bemerkung zu Henkin's Beweis für die Vollständigkeit des Prädikatenkalküls der ersten Stufe.* Journal of Symbolic Logic, 18(1953), 42–48.

[13] HENKIN L.: *The completeness of the first-order functional calculus.* Journal of Symbolic Logic, 14(1951), 159–166.

[14] HENKIN L.: *La structure algébrique des théories mathématiques.* Paris 1956.

[15] HENKIN L., TARSKI A.: *Cylindric algebras.* Proceedings of Symposia in Pure Mathematics II(1961), Lattice theory, 83–113.

[16] HERBRAND J.: *Recherches sur la théorie de la démonstration.* Travaux de la Société des Sciences et des Lettres de Varsovie, Classe III, 33(1930).

[17] HILBERT D., ACKERMANN W.: *Grundzüge der theoretischen Logik.* Berlin 1928.

[18] HILBERT D., BERNAYS P.: *Grundlagen der Mathematik.* Vol. II. Berlin 1939.

[19] HINTIKKA K. J. J.: *Form and content in quantification theory.* Acta Philosophica Fennica, 8(1955), 7–55.

[20] KEISLER H. J.: *A survey of ultraproducts.* In: Logic, methodology and philosophy of sciences. Amsterdam 1965, 112–126.

[21] KLEENE S. C.: *Introduction to metamathematics.* Amsterdam–Groningen 1952.

[22] LYNDON R. C.: *Notes on logic.* New York 1966.

[23] Łoś J.: *The algebraic treatment of the methodology of elementary deductive systems.* Studia Logica, 2(1955), 151–212.

[24] Łoś J.: *Quelques remarques, théorèmes et problèmes sur les classes définissables d'algèbres.* In: Mathematical interpretations of formal systems. Amsterdam 1955, 98–113.

[25] MacNeille H.: *Partially ordered sets.* Transactions of the American Mathematical Society, 42(1937), 416–460.

[26] Mal'cev A. I.: *Untersuchungen aus dem Gebiete der mathematischen Logik.* Matematičeskij Sbornik, 1(43)(1936), 323–336.

[27] Mal'cev A. I.: *Ob odnom obščem metode polučenija lokal'nyh teorem teorii grupp.* Učenyje Zapiski Iwanowskogo Pedagogičeskogo Instituta. Fiziko-Matematičeskij Fakultet, 1(1941), 3–9.

[28] Mendelson E.: *Introduction to mathematical logic.* New York 1964.

[29] Morel A., Scott D., Tarski A.: *Reduced products and the compactness theorem.* Notices of the American Mathematical Society, 5(1958), 674.

[30] Mostowski A.: *Logika matematyczna.* Warszawa–Wrocław 1948.

[31] Mostowski A.: *Sur l'interprétation géometrique et topologique des notions logiques.* Proceedings of the Tenth International Congress of Philosophy (Amsterdam, August 11–18, 1948), 1(1949), 767–769.

[32] Pogorzelski W. A.: *Klasyczny rachunek zdań. Zarys teorii.* Warszawa 1969.

[33] Rasiowa H.: *A proof of the compactness theorem for arithmetical classes.* Fundamenta Mathematicae, 39(1952), 8–14.

[34] Rasiowa H., Sikorski R.: *A proof of the completeness theorem of Gödel.* Fundamenta Mathematicae, 37(1950), 193–200.

[35] Rasiowa H., Sikorski R.: *The mathematics of metamathematics.* Warszawa 1968.

[36] Reichbach J.: *On the completeness of the functional calculus of first order.* Studia Logica, 2(1955), 245–250.

[37] Rieger L.: *On countable generalized σ-algebras, with a new proof of Gödel's completeness theorem.* Czechoslowak Mathematical Journal, 1(76) (1951), 29–40.

[38] Rieger L.: *On free $\aleph_\xi$-complete Boolean algebras.* Fundamenta Mathematicae, 38(1951), 35–52.

[39] Rieger L.: *Algebraic methods of mathematical logic.* Prague 1967.

[40] Robinson A.: *On the construction of models.* In: Essays on the foundations of mathematics. Jerusalem 1961, 207–217.

[41] Robinson A.: *Recent developments in model theory.* In: Logic, methodology and philosophy of science. Proceedings of the 1960 International Congress, 1962, 60–79, Stanford.

[42] Robinson A.: *Introduction to model theory and to the metamathematics of algebra.* Amsterdam 1963.

[43] Sikorski R.: *Boolean algebras.* Third edition. Berlin 1969.

[44] Skolem T.: *Logisch-kombinatorische Untersuchungen über die Erfüllbarkeit oder Beweisbarkeit mathematischer Sätze nebst einem Theoreme über dichte Mengen.*

Videnskapsselskapets Skrifter, I. Matematisk-naturvidenskabelig klasse, 4(1920).

[45] Skolem T.: *Über die Nicht-charakterisierbarkeit der Zahlenreihe mittels endlich oder abzählbar unendlich vieler Aussagen mit ausschliesslich Zahlenvariablen.* Fundamenta Mathematicae, 23(1934), 150–161.

[46] Słupecki J.: *Funkcja Łukasiewicza.* Zeszyty Naukowe Uniwersytetu Wrocławskiego, seria B, 3 (1959), 33–40.

[47] Słupecki J., Pogorzelski W. A.: *A variant of the proof of the completeness of the first order functional calculus.* Studia Logica, 12 (1961), 125–134.

[48] Smullyan R.: *First-order logic.* Berlin 1968.

[49] Tarski A.: *Une contribution à la théorie de la mesure.* Fundamenta Mathematicae, 15 (1930), 42–50.

[50] Tarski A.: *Über einige fundamentale Begriffe der Metamathematik.* Comptes Rendus des Séances de la Société des Sciences et des Lettres de Varsovie, Classe III, 23(1930), 22–29.

[51] Tarski A.: *Some notions and methods on the borderline of algebra and metamathematics.* Proceedings of the International Congress of Mathematicians, Cambridge, Massachusetts, U.S.A., 1950, vol. 1, 1952, 705–720, Providence.

[52] Wang H.: *A survey of mathematical logic.* Amsterdam 1962.

[53] Zygmunt J.: *Kurt Gödel's doctoral dissertation.* This volume.

STANISŁAW J. SURMA

THE CONCEPT OF THE LINDENBAUM ALGEBRA: ITS GENESIS [1]

1. The method of the Lindenbaum algebra was worked-out in 1926–1927. It was acknowledged and applied again in the domain of the propositional calculi in the forties and in predicate calculi in the late forties and the early fifties. I shall outline here the genesis of this method and some of its generalizations with special applications to the logical research while omitting many other applications of this method as well as an interesting relationship between Lindenbaum's construction and the well-known Gödel's arithmetization.

2. The original idea of the discussed method of the Lindenbaum algebra was first used by Adolf Lindenbaum, a young but prominent mathematician of Warsaw University who was killed by German fascists during the Second World War. His idea consists in constructing logical matrices for the propositional calculus with formulas of the formalized language of the calculus under consideration. It was first devised in the academic year 1926–1927 at Professor Jan Łukasiewicz's seminar in mathematical logic held in Warsaw University. Lindenbaum used his idea to prove the following

THEOREM 1. *For any set of sentences X expressed in the language of the propositional calculus and closed under substitution for the propositional variables and under detachment there exists at most denumerable normal matrix M such that X is equal to the content of M* [2].

[1] This article contains the text of a lecture given by the author on April 29, 1967 to the XIIIth Conference of the History of Logic organized in Cracow by the Section of Logic, Polish Academy of Sciences, together with the Department of Logic of the Jagiellonian University.

An earlier version of this article was published under the title „*History of logical applications of the method of Lindenbaum's algebra*" in Acta Logica, 10 (1967), 127–138, Analele Universitatii Bucuresti.

[2] The term „normal matrix" will be defined later.

Recently, theorem 1 is called the Lindenbaum theorem of existing adequate normal matrices and is treated as the most fundamental theorem in the theory of the logical matrices. It was published without proof in 1930 in a joint paper by J. Łukasiewicz and A. Tarski [13].

Let us note that in theorem 1 one can omit altogether the conditions stating that X is closed under detachment and that M is normal.

The idea of treating the set of sentences of a formalized language as a logical matrix was presented again by Lindenbaum in his lecture given at the first Polish Congress of Mathematicians, held in Lwów, in September 1927. Unfortunately the summary of this lecture contains no reference to the mentioned idea (cf. [10]).

Thus, the Lindenbaum's method has never been published by its author but has remained in the oral tradition of the Warsaw Mathematical School. One must note, however, that in the Warsaw School in the thirties the proper significance of the Lindenbaum's method as a tool for constructing logical matrices was lost though some mathematicians were trained back in this period to similar ideas. Thus in 1935 Tarski introduced an algebra of sentences with operations corresponding to logical connectives (cf. [29]). Apart from a different treatment of the notion of equality in this algebra, it coincided with Lindenbaum's construction. One can also observe the close relationship between the Lindenbaum's method and the method of proof of the completeness theorem for the propositional calculi presented in 1936 by M. Wajsberg in [31] (cf. also [1]) since the Wajsberg's method consists in constructing a set of the so-called logical values of the corresponding sentences of the propositional calculus under consideration.

The Lindenbaum's method came to be known abroad through A. Tarski, one of the most active participants in Łukasiewicz's seminar in the twenties and thirties. It was first published by J. C. C. McKinsey, a close collaborator of Tarski, in his paper [16] in 1941. McKinsey called this method „an unpublished method of Lindenbaum“ „explained to me by Professor Tarski“. In [16] a proof of a particular case of theorem 1 is contained. But in this proof one can find the principal idea of the original proof of theorem 1 devised by Lindenbaum.

In 1949 the paper [11] of Jerzy Łoś was published. This paper is a large and systematic study of the Lindenbaum method. It describes the method of constructing logical matrix of the sentences of any propositional calculus, called by Łoś m-th Lindenbaum's matrix of meaningful expressions, and presents a reduction of this matrix by suitable congruence relation defined in a set of all sentences. This congruence is called by Łoś the interexchangeability relation on the basis of fixed set of sentences and the matrix reduced by this congruence is called the m-th Lindenbaum matrix of proofs. Let us note that Łoś replaces each equivalence class of interexchangeability by a suitable element of this class called the representative of this class. In the study [11] a detailed proof of theorem 1 is contained.

Below we are formulating exact definitions and the fundamental properties of these matrices.

By S we denote the least set including propositional variables p_i (where $i \in N$) closed under k_i-argument connectives Φ_i (where $i \in I_0 \subset N$). Thus if $A_1, A_2, \ldots, A_{k_i} \in S$, then $\Phi_i(A_1, A_2, \ldots, A_{k_i}) \in S$. By S_m (where $m \leqslant \aleph_0$) we denote the set of sentences containing no propositional variables apart from $p_0, p_1, \ldots, p_{m-1}$. It is obvious that $S_m \subset S_{m+1}$ and $S_{\aleph_0} = S$.

Here the symbols $\in$, $\subset$, N and $\aleph_0$ denote the membership relation, the inclusion relation, the set of all natural numbers and the cardinal (number of elements) of any denumerably infinite set, respectively.

By M we denote the logical matrix, i.e., the sequence

$$\langle W, W^*, \{\varphi_i\}_{i \in I_0} \rangle$$

where (i) W is a non-empty set of the so called logical values assumed to be the domain of ranging of the propositional variables,

(ii) W^* is a subset of W consisting of the so called distinguished logical values and

(iii) φ_i is a k_i-argument operation in the set W corresponding to the connective Φ_i.

The normal matrix is, by definition, the matrix in which the operation φ corresponding to the implication connective fulfils the condition stating that for any $w, v \in W$, if $w \in W^*$ and $v \in W \setminus W^*$, then $\varphi(w, v) \in W \setminus W^*$.

Here the symbol $\setminus$ denotes the set-theoretical operation of difference.

By the cardinal of a matrix M we understand the cardinal of a set W.

EXAMPLE 1. *Let us suppose that* $W = \{0, 1\}$, $W^* = \{1\}$, $I_0 = \{1, 2\}$, $k_1 = 2$, $k_2 = 1$, φ_1 *corresponds to the implication* Φ_1 *denoted here by* $\rightarrow$, *and* φ_2 *corresponds to the negation* Φ_2 *denoted by* $\neg$. *If* $\varphi_1(1, 1) = \varphi_1(0, 1) = \varphi_1(0, 0) = \varphi_2(0)$ *and* $\varphi_1(1, 0) = \varphi_2(1) = 0$, *then* M *is called the classical two-valued logical matrix and is denoted by* $M^{(2)}$.

By $E(M)$ we denote the content of the matrix M, i.e., the set of sentences obtaining the elements of the set W^* as their values under the condition that all the variables and connectives Φ_i occurring in these sentences are replaced by any elements of W and by the operations φ_i, respectively.

M is said to be the m-th Lindenbaum's matrix of the meaningful expressions determined by the sets $X \subset S$ and S_m, in symbols, $M = M(X, S_m)$, if and only if

(i) X is closed under substitution,

(ii) $W = S_m$,

(iii) $W^* = X \cdot S_m$ and

(iv) $\varphi_i(A_1, A_2, \ldots, A_{k_i}) = \Phi_i(A_1, A_2, \ldots, A_{k_i})$ for any $A_1, A_2, \ldots, A_{k_i} \in S_m$.

One obtains the following

LEMMA 1. $S_m \cdot E(M(X, S_m)) = S_m \cdot X$.

Hence it follows immediately for $m = \aleph_0$

COROLLARY. $E(M(X, S)) = X$.

The sentences A and B are said to be interexchangeable on the basis of $X \subset S$, in symbols, $A \sim_X B$, if and only if

(i) X is closed under substitution and

(ii) for any $C(A) \in S$ the condition stating that $C(A) \in X$ is equivalent to the condition stating that $C(A/B) \in X$ where $C(A)$ is a sentence containing as a part a sentence A and where $C(A/B)$ is a sentence resulting from a sentence $C(A)$ by substitution of B for A.

The relation $\sim_X$ is a congruence in S. Let us observe that if X is the set of all the theorems of the classical propositional calculus with the implication connective $\rightarrow$, then the condition stating that $A \sim_X B$ is equivalent to the condition stating that $A \rightarrow B \in X$ and $B \rightarrow A \in X$.

In the sequel instead of $\sim_X$ we shall write simply $\sim$. The equivalence class determined by $\sim$ and A is denoted by $[A]_\sim$ or simply by $[A]$ in the cases when the relation $\sim$ is known. The reduction of a set Z by relation $\sim$ is denoted here by $Z/_\sim$.

M is said to be the m-th Lindenbaum matrix of proofs determined by sets $X \subset S$ and S_m, in symbols, $M = M(X, S_m)/_\sim$, if and only if

(i) X is closed under substitution,

(ii) $W = S_m/_\sim$,

(iii) $W^* = (X \cdot S_m)/_\sim$ and

(iv) $\varphi_i([A_1], [A_2], \ldots, [A_{k_i}]) = [\Phi_i(A_1, A_2, \ldots, A_{k_i})]$

for any $A_1, A_2, \ldots, A_{k_i} \in S$.

One obtains the following

LEMMA 2. $E(M(X, S_m)/_\sim) = E(M(X, S_m))$.

EXAMPLE 2. *Let us suppose that* $I_0 = \{1, 2, 3\}$, $k_1 = k_2 = 2$, $k_3 = 1$, Φ_1 *is the disjunction connective* $\vee$, Φ_2 *is the conjunction connective* $\wedge$, Φ_3 *is the negation connective* $\rightharpoondown$ *of the classical propositional calculus and let us denote by* $+$, $\cdot$ *and by* $-$ *the operations* φ_1, φ_2 *and* φ_3 *corresponding to* Φ_1, Φ_2 *and* Φ_3, *respectively. Then the m-th Lindenbaum matrix of proofs determined by the set of all the theorems of the classical propositional calculus with disjunction, conjunction and negation and by the set* S_m *satisfies the following equalities*

$$[A]+[B] = [A \vee B], \qquad [A]\cdot[B] = [A \wedge B]$$

and

$$S_m/_{\sim}\backslash[A] = [\rightharpoondown A]$$

and forms a Boolean algebra with operations $+$, $\cdot$ *and* $\backslash$. *The unit element of this algebra is the set of all the theorems of the classical propositional calculus expressed by means of the propositional variables* $p_1, p_2, \ldots, p_m$. *Let us observe that the Boolean interpretation of the implication has the form*

$$[A \rightarrow B] = (S_m/_{\sim}\backslash[A]) + [B].$$

The described algebra is isomorphic to $M^{(2)}$.

The terminology used in [11] was abandoned because of its artificiality. In particular, the matrix $M(X, S)/_{\sim}$ is called in the recent literature the Lindenbaum algebra (cf., e.g., [18]) or the Lindenbaum-Tarski algebra (cf., e.g., [21] and [9]).

3. The original Lindenbaum's method was applied to the propositional calculi. In the late forties and in the early fifties some wider generalizations of Lindenbaum's construction were obtained and then applied in the first order predicate calculus, especially to prove the following Gödel-Malcev completeness theorem (cf. [4] and [14]):

THEOREM 2. *For every consistent set of sentences* X *expressed in a language of first order predicate calculus there exists a model* $\mathfrak{M}$ *verifying* X.

These proofs were based, as a rule, on some modifications of the following well-known theorem on deductively complete systems:

THEOREM 3. *For any consistent set of sentences* X *expressed in a language of first order predicate calculus there exists a consistent and deductively complete system* Y *closed under detachment and including* X.

Let us recall that X is said to be consistent, if and only if, a sentence A such that both A and $\rightharpoondown A$ are theorems in X does not exist, and X

is said to be deductively complete, if and only if for any sentence A at least one holds: A or $\rightarrow A$ is a theorem in X.

Theorem 3 was formulated and proved by Lindenbaum but not published by its author. It was published without proof in 1930 in Tarski's paper [26].

In the sequel we present three independent generalizations of the Lindenbaum's method. The first of them is due to J. Łoś, the second is due to L. Henkin and others, and the third to H. Rasiowa and R. Sikorski and also to L. Rieger.

In [11] Łoś constructed the so-called algebraic Lindenbaum matrix of proofs applying it to prove theorem 2 for the open systems, i.e., for the systems consisting of the sentences without bound variables. We are now formulating an exact definition of the mentioned algebraic matrix.

By T we denote the least set containing the set of the individual variable symbols z_i (where $i \in N$) and closed under m_i-argument functors F_i (where $i \in I_1 \subset N$). The elements of T are named terms. By S_T we denote the least set containing

(i) propositional variables p_i,

(ii) elementary sentences or predicates $R_i(t_1, t_2, \ldots, t_{n_i})$ (where R_i is n_i-argument predicate symbol, $i \in I_2 \subset N$ and $t_1, t_2, \ldots, t_{n_i} \in T$) and

(iii) closed under connectives Φ_i.

Obviously, $S \subset S_T$.

By T_n (where $n \leqslant \aleph_0$) we denote the set of terms containing no individual variables apart from $z_0, z_1, \ldots, z_{n-1}$. By $S_{m,n}$ (where $m \leqslant \aleph_0$ and $n \leqslant \aleph_0$) we denote the set of sentences containing no propositional variables apart from $p_0, p_1, \ldots, p_{m-1}$ and individual variables $z_0, z_1, \ldots$ $\ldots, z_{n-1}$. Let us observe that $S_{k,l} \subset S_{m,n}$ for $k \leqslant m$ and $l \leqslant n$, *and* $S_{\aleph_0,\aleph_0} = S_T$.

A sequence $\mathfrak{M} = \langle M, \mu \rangle$ is said to be an algebraic matrix, if $M = \langle W, W^*, \{\varphi_i\}_{i \in I_0} \rangle$ is a logical matrix and $\mu = \langle U, \{f_i\}_{i \in I_1}, \{r_i\}_{i \in I_2} \rangle$ is an algebraic-relational structure, i.e., the structure in which

(i) U is an arbitrary non-empty set (the so-called universe of the individuals) disjoint with W and assumed to be the domain of ranging of the individual variable symbols,

(ii) f_i is the m_i-argument operation in U corresponding to the functor F_i and

(iii) r_i is the n_i-argument relation in U, i.e., the n_i-argument function from U to W corresponding to the predicate symbol R_i.

By the cardinal of an algebraic matrix $\mathfrak{M}$ we understand the cardinal of the set U.

The content of the algebraic matrix $\mathfrak{M}$, i.e., the set of the sentences $E(\mathfrak{M})$ is defined by the well-known Tarski's method (cf. [28]).

We are now modifying the interexchangeability relation as follows. Two sentences A and B are said to be interexchangeable on the basis of $X \subset S_T$, if and only if

(i) X is closed under substitution and

(ii) for any $C(A) \in S_T$ the condition stating that $C(A) \in X$ is equivalent to the condition stating that $C(A/B) \in X$ where A and B are the sentences or the terms, respectively.

This relation is a congruence in S_T or in T respectively. In the sequel we shall denote it by $\approx$ and, as a rule, we omit it in the symbol $[A]_\approx$.

$\mathfrak{M}$ is said to be the (m, n)-th algebraic Lindenbaum's matrix of proofs determined by the sets $X \subset S_T$, $S_{m,n}$ and T_n, in symbols, $\mathfrak{M} = \mathfrak{M}(X, S_{m,n}, T_n)/_\approx$, if and only if

(i) X is closed under substitution,

(ii) the sets X and $S_{m,n}$ determine the m-th Lindenbaum matrix of proofs $M(X, S_{m,n})/_\approx$,

(iii) $U = T_n/_\approx$,

(iv) $f_i([t_1], [t_2], \ldots, [t_{m_i}]) = [F_i(t_1, t_2, \ldots, t_{m_i})]$, for any $t_1, t_2, \ldots, t_{m_i} \in T_n$ and

(v) $r_i([t_1], [t_2], \ldots, [t_{n_i}])$ holds in U if and only if $R_i(t_1, t_2, \ldots, t_{n_i}) \in X$, for any $t_1, t_2, \ldots, t_{n_i} \in T$.

This definition can be simplified as follows. Assuming that no two different elements of T are mutually interexchangeable or assuming that the functors F_i are zero-argument (i.e., $m_i = 0$, for any $i \in I_1$) we are enabled to write (i) and (ii) as above, and further,

(iii) $U = T_n$,

(iv) is vanishing now, and

(v) $r_i(t_1, t_2, \ldots, t_{n_i})$ holds in U, if and only if $R_i(t_1, t_2, \ldots, t_{n_i}) \in X$, for any $t_1, t_2, \ldots, t_{n_i} \in T_n$.

One obtains the following

LEMMA 3. $S_{m,n} \cdot E(\mathfrak{M}(X, S_{m,n}, T_n)/_\approx) = S_{m,n} \cdot X$.

In Łoś [11] theorem 2 was proved for open sentences expressed in a language of the first order predicate calculus basing on the following lemmas

LEMMA 4. *For any set of open sentences X closed under substitution and detachment and such that $X\cdot S = E(M^{(2)})$ there exists a maximal set Y including X, closed under substitution and detachment and such that $Y\cdot S = E(M^{(2)})$.*

LEMMA 5. *If X is a maximal set of open sentences closed under substitution and detachment and such that $X\cdot S = E(M^{(2)})$, then there exist numbers m, n such that $1 \leqslant m, n \leqslant \aleph_0$ and $X = E\big(\mathfrak{M}(X, S_{m,n}, T_n)/_{\approx}\big)$.*

Lemma 4 is a modification of theorem 3 whereas lemma 5 is a modification of theorem 1.

By making use of the well-known Skolem method of eliminating quantifiers, the above method of proving theorem 2 can be applied also to the systems of closed sentences, i.e., sentences without free variables. Namely, let X be a set of sentences of the first order language. Let us denote by $Pr(X)$ the set of all the prenex normal form of all the elements of X. Let $G = \{G_i\}_{i\in I_3}$ be a set of functors such that $I_3 \neq I_j$ for $j = 0, 1, 2$ and let $Sk_G\big(Pr(X)\big)$ denotes Skolem reduction of the set $Pr(X)$, i.e., the set of sentences resulting from $Pr(X)$ by Skolem method of eliminating quantifiers each of this sentences being generalized by universal quantifiers. From the above it follows readily that the set $Sk\big(Pr(X)\big)$ is consistent if only the set X is consistent and that $X \subset E(\mathfrak{M})$, if $Sk_G\big(Pr(X)\big) \subset E(\langle\mathfrak{M}, \{g_i\}_{i\in I_3}\rangle)$, where g_i are operations in U corresponding to the functors G_i.

This manner of proving theorem 2 was presented by Łoś in his later paper [12] prepared to print in 1953 but published in 1955. In [12] Łoś informs that the mentioned method was presented to the Wrocław Section of the Polish Mathematical Society on January 12, 1951. Let us observe that quite a similar manner of proving theorem 2 we find in E. W. Beth's paper [2] published in 1951.

The algebraic matrices in Łoś' terminology, i.e., sequences of the form $\mathfrak{M} = \langle M, \mu\rangle$ are called recently the algebraic models or simply models (cf. [19]).

Henkin's paper [7] was prepared to print in 1948 and published in January 1949. It contains some generalization of Lindenbaum's method with application to proving theorem 2. In the sequel we describe the Henkin's method in details.

Henkin first describes a denumerable sequence of inductions using as a fact theorem 3. Namely, let S_0 be a set of closed sentences (without constant individual symbols) expressed in the classical first order language. This set can be arranged, e.g., in the form

$$S_0 = \{A_{0,1}, A_{0,2}, A_{0,3}, \ldots\}.$$

Let $X \subset S_0$. Let us suppose that X is consistent and assume that

(i) $Y_{0,0} = X$,

(ii) $Y_{0,j+1} = Y_{0,j} + \{A_{0,j+1}\}$, if $Y_{0,j} + \{A_{0,j+1}\}$ is consistent,

(iii) $Y_{0,j+1} = Y_{0,j}$ otherwise

and

(iv) $Y_0 = \bigcup_{j=0}^{\infty} Y_{0,j}$.

Here the symbol $\bigcup_{j=0}^{\infty} X_j$ denotes the sum $X_0 + X_1 + X_2 + \ldots$ It is easy to verify that $Y_0 \subset S_0$ and that Y_0 is a consistent and complete set closed under detachment. Let us arrange the elements of Y_0 like that

$$Y_0 = \{A_{0,1}, A_{0,2}, \ldots\}$$

and let $\{t_{1,1}, t_{1,2}, t_{1,3}, \ldots\}$ be a sequence of the constant individual symbols (zero-argument functors) not occurring in the elements of Y_0. Let us denote by S_1 the set of all sentences of the language extended in this way. Let us assume that

(v) $X_{1,0} = Y_0$,

(vi) $X_{1,j+1} = X_{1,j} + \{B_{1,j}(x/t_{1,j})\}$, if there exists a sentence $B_{1,j}(x)$ with one free variable x such that $A_{0,j}$ is equiform with $\bigvee x B_{1,j}(x)$,

(vii) $X_{1,j+1} = X_{1,j}$ otherwise

and

(viii) $X_1 = \bigcup_{j=0}^{\infty} X_{1,j}$.

Here the symbol $\bigvee x$ denotes the existential quantifier binding the variable x. Let us arrange the elements of S_1 like that

$$S_1 = \{A_{1,1}, A_{1,2}, A_{1,3}, \ldots\}$$

and let us assume that

(ix) $Y_{1,0} = X_1$,

(x) $Y_{1,j+1} = Y_{1,j} + \{A_{1,j+1}\}$, if $Y_{1,j} + \{A_{1,j+1}\}$ is consistent,

(xi) $Y_{1,j+1} = Y_{1,j}$ otherwise

and

(xii) $Y_1 = \bigcup_{j=0}^{\infty} Y_{1,j}$.

It is easy to verify that Y_1 is a consistent and complete set closed under detachment.

Now let us assume inductively that

$$Y_i = \{A_{i,1}, A_{i,2}, A_{i,3}, \ldots\} \subset S_i$$

where S_i results from S_{i-1} by adding the new constant individual symbols $\{t_{i,j}\}_{j=1,2,3,\ldots}$ not occurring in the elements of Y_i and let us suppose that Y_i is a consistent and complete set closed under detachment. With the help of $\{t_{i+1,j}\}_{j=1,2,3,\ldots}$ we form the set S_{i+1} and we assume that

(xiii) $X_{i+1,0} = Y_i$,

(xiv) $X_{i+1,j+1} = X_{i+1,j} + \{B_{i+1,j}(x/t_{i+1,j})\}$ if there exists a sentence $B_{i+1,j}(x)$ with one free variable x such that $A_{i,j}$ is equiform with $\bigvee x B_{i+1,j}(x)$,

(xv) $X_{i+1,j+1} = X_{i+1,j}$ otherwise

and

(xvi) $X_{i+1} = \bigcup_{j=0}^{\infty} X_{i+1,j}$.

We arrange the elements of S_{i+1} like that

$$S_{i+1} = \{A_{i+1,1}, A_{i+1,2}, A_{i+1,3}, \ldots\}$$

and put

(xvii) $Y_{i+1,0} = X_{i+1}$,

(xviii) $Y_{i+1,j+1} = Y_{i+1} + \{A_{i+1,j+1}\}$, if $Y_{i+1,j} + \{A_{i+1,j+1}\}$ is consistent,

(xix) $Y_{i+1,j+1} = Y_{i+1,j}$ otherwise

and

(xx) $Y_{i+1} = \bigcup_{j=0}^{\infty} Y_{i+1,j}$.

Obviously, $Y_{i+1} \subset S_{i+1}$ and Y_{i+1} is a consistent and complete set closed under detachment.

Finally we put

(xxi) $S_\omega = \bigcup_{i=0}^{\infty} S_i$,

(xxii) $Y_\omega = \bigcup_{i=0}^{\infty} Y_i$

and

(xxiii) $T_{\omega,\omega} = \{t_{i,j}\}_{i,j=1,2,3,\ldots}$

The following lemma can be proved.

LEMMA 6. *Y_ω is a consistent and complete set closed under detachment and such that for any sentence $A(x)$ with one free variable x if $\bigvee x A(x) \in Y_\omega$, then there exists $t \in T_{\omega,\omega}$ such that $A(x/t) \in Y_\omega$.*

The above construction is simple because of assuming that

(i) in the elements of S_0 no constant individual symbols occur,

(ii) we add as new symbols the constant individual symbols instead of constant terms and

(iii) the number j is an integer.

Of course, all these assumptions can be abandoned. In particular, one can put that $j < \gamma_0$, where γ_0 is any transfinite ordinal, obtaining thus T_{ω,γ_0} instead of $T_{\omega,\omega}$. In this last case the Zermelo's axiom of choice must be used.

Now Henkin describes a model which is essentially the algebraic Lindenbaum's model $\mathfrak{M}(Y_\omega, S_\omega, T_{\omega,\omega})$ though he does not make any explicit references to Lindenbaum himself and he shows that the following lemma holds

LEMMA 7. $Y_\omega = E(\mathfrak{M}(Y_\omega, S_\omega, T_{\omega,\omega}))$.

This completes the proof of theorem 2 by Henkin's method in view of $X \subset Y_\omega$.

The above Henkin's method was applied to classical predicate calculus. Later in 1950 in paper [8] it was extended, so as to be also applicable to some of the non-classical predicate calculi.

Among the logicians Henkin's method is well-known and it was simplified and modified by many authors. We describe here the simplification due to G. Hasenjaeger (cf. [6]).

Let S_0 be the set of all closed sentences (without constant individual symbols) expressed in a classical first order language and let $X \subset S_0$. Let us assume that X is a consistent set. Let us consider the set $T_\omega = \{t_1, t_2, t_3, \ldots\}$ of constant individual symbols and let us denote by S_1 the set of sentences constructed with the help of T_ω. We denote by X^* the set resulting from X by adding to it the logical axioms expressed in the language involving the new constants. Of course, $X^* \subset S_1$ and it is easy to verify that X^* is a consistent set.

Let $\{A_k(x_{i,k})\}_{k=1,2,3,\ldots}$ be the set of all sentences having one free variable $x_{i,k}$. Let us choose a sequence $\{t_{j,1}, t_{j,2}, t_{j,3}, \ldots\} \subset T_\omega$ such that $t_{j,k}$ occurs in none of the formulas $A_1(x_{i,1}), A_2(x_{i,2}), \ldots, A_k(x_{i,k})$, for $k = 1, 2, \ldots$ and such that $t_{j,k}$ is different from each of $t_{j,1}, t_{j,2}, \ldots$ $\ldots, t_{j,k-1}$. Let us denote by B_k the sentence

$$\bigvee x_{i,k} A_k(x_{i,k}) \to A_k(x_{i,k}/t_{j,k})$$

and let us put

(i) $X_{1,0} = X^*$,

(ii) $X_{1,i+1} = X_{1,i} + \{B_{i+1}\}$

and

(iii) $X_1 = \bigcup_{i=0}^{\infty} X_{1,i}$.

It is easy to verify that $X_1 \subset S_1$ and that X_1 is a consistent set closed under detachment. Finally let us put

(iv) $Y_{1,0} = X_1$,

(v) $Y_{1,i+1} = Y_{1,i} + \{A_{i+1}\}$, if $Y_{1,i} + \{A_{i+1}\}$ is consistent

(vi) $Y_{1,i+1} = Y_{1,i}$ otherwise

and

(vii) $Y_1 = \bigcup_{i=0}^{\infty} Y_{1,i}$.

Of course, $Y_1 \subset S_1$ and there holds the lemma resulting from lemma 6 by replacing Y_ω and $T_{\omega,\omega}$ by Y_1 and T_ω, respectively.

Let us note that the variants of Henkin-Hasenjaeger's method are discussed also by A. Grzegorczyk in [5] as well as in the papers [20], [24] and [25].

The paper [18] of H. Rasiowa and R. Sikorski was published in 1950 and it was concerned with the proof of theorem 2 for the systems of closed sentences based on the classical first order predicate calculus. The detailed exposition of authors' method is contained in their monograph [19] published in 1963. In the below description of this method we use the notation used above while presenting the method of Henkin-Hasenjaeger.

$\mathfrak{M}$ is said to be the quantifier algebraic Lindenbaum's model determined by $X \subset S_1$ and T_ω, if and only if $\mathfrak{M} = \langle M, \mu, \bigcup \rangle$ where

(i) M is the matrix described in example 2,

(ii) μ is an algebraic-relational structure described by Henkin and Hasenjaeger and

(iii) $\bigcup$ is a function from T_ω to the set of all the mappings of the set $S_1/_{\approx}$ into itself satisfying the condition stating that $\bigcup_{t \in T_\omega} [A(x/t)] = [\bigvee x A(x)]$, for any sentence $A(x)$ with one free variable x.

In the above definition the existential quantifier is interpreted as an operation of infinite meet in the (complete) Boolean algebra. Of course, this interpretation is not original. We find it in an earlier paper [15] as well as in another important paper [17].

In proving theorem 2 Rasiowa and Sikorski use the two auxiliary lemmas. The first of them is a modification of the well-known theorem of the extension of ideals to prime ideals. This last theorem has been formulated and proved separately by several authors and for different purposes (cf. [30], [27], [23]). It is closely connected with theorem 3 for

Lindenbaum algebras determined, respectively, by consistent or by consistent and complete sets of sentences expressed in the classical first order language form Boolean ideals and Boolean prime ideals, respectively. Rasiowa and Sikorski use in [18] the following modification of the theorem of extension of ideals to prime ideals.

LEMMA 8. *If*

(i) a_0 *is an element of a Boolean algebra* $\mathfrak{A}$ *different from the unit element and*

(ii) *for any* $i \in I$ *if* $a_i \in \mathfrak{A}$, *then* $\bigcup_{i \in I} a_i \in \mathfrak{A}$,

then there exists in $\mathfrak{A}$ *a prime ideal* J *such that* $a_0 \in J$ *and* $[\bigcup_{i \in I} a_i]_J = \bigcup_{i \in I} [a_i]_J$, *where the symbol* $[a]_J$ *denotes the equivalence class determined by the element* a *and by a congruence induced by the prime ideal* J.

Let us observe that a variant of a simplified proof of lemma 8 is due to Tarski (cf. [3]).

The second lemma which is needed for the proof of theorem 2 by Rasiowa-Sikorski's method can be formulated as follows.

LEMMA 9. *If*

(i) $X \subset S_1$,

(ii) $X/_{\approx}$ *is a prime ideal in* $S_1/_{\approx}$,

(iii) *for any sentence* $A(x)$ *with one free variable* x *the equality*

$$\Big[\bigcup_{t \in T_\omega} [A(x/t)]_{\approx}\Big]_{X/\approx} = \bigcup_{t \in T_\omega} \big[[A(x/t)]_{\approx}\big]_{X/\approx}$$

holds and

(iv) $[a_0]_{\approx} \in X/_{\approx}$,

then $a_0 \notin E(\mathfrak{M}(X, S_1, T_\omega))$.

Quite a similar method of proving theorem 2 has been developed by L. Rieger in [21] and [22] in 1951. In particular, Rieger proves the theorem on representation of the quantifier Lindenbaum algebra in a denumerable field of all Borel subset of the Cantor discontinuum. From this theorem theorem 2 follows immediately. Let us observe that according to Rieger's information contained in [21] his method has been presented by its author at Professor A. Mostowski's seminar in mathematical logic in March and April 1950 and at the session of Wrocław Section of the Polish Mathematical Society in May 1950.

The above considerations show that all the well-known recent methods of proving Gödel-Malcev completeness theorem are based on the same ideas, namely, on the important ideas due to A. Lindenbaum.

References

[1] ASSER G.: *Einführung in die mathematische Logik*. Teil I.
Leipzig 1959.

[2] BETH E. W.: *A topological proof of the theorem of Lindenbaum-Skolem-Gödel-Tarski.*
Proceedings of the Royal Academy of Sciences, series A, 54(1951), 436–444, Amsterdam.

[3] FEFERMAN S.: *Review of the paper: H. Rasiowa and R. Sikorski, „A proof of the completeness theorem of Gödel“.*
Fundamenta Mathematicae, 37(1950), 193–200.
Journal of Symbolic Logic, 19(1952), 72.

[4] GÖDEL K.: *Die Vollständigkeit der Axiome der logischen Funktionenkalkuls.*
Monatshefte für Mathematik und Physik, 37(1930), 349–360.

[5] GRZEGORCZYK A.: *Zarys logiki matematycznej*. Wydanie drugie.
Warszawa 1969.

[6] HASENJAEGER G.: *Bemerkung zu Henkin's Beweis für die Vollständigkeit des Prädikatenkalküls der ersten Stufe.*
Journal of Symbolic Logic, 18(1953), 42–48.

[7] HENKIN L.: *A proof of completeness for the first order functional calculus.*
Ibid., 14(1949), 159–166.

[8] HENKIN L.: *An algebraic characterization of quantifiers.*
Fundamenta Mathematicae, 37(1950), 63–74.

[9] HENKIN L., TARSKI A.: *Cylindric algebras.*
Proceedings of Symposia in Pure Mathematics II (1961), Lattice theory, 83–113.

[10] LINDENBAUM A.: *Méthodes mathématiques dans les recherches sur le système de la théorie de déduction.*
Księga Pamiątkowa Pierwszego Polskiego Zjazdu Matematycznego, Lwów, 7–10 IX 1927.
Kraków 1929.

[11] ŁOŚ J.: *O matrycach logicznych.*
Prace Wrocławskiego Towarzystwa Naukowego, Seria B, Nr 19.
Wrocław 1949.

[12] ŁOŚ J.: *The algebraic treatment of the methodology of elementary deductive systems.*
Studia Logica, 2(1955), 151–212.

[13] ŁUKASIEWICZ J., TARSKI A.: *Untersuchungen über den Aussagenkalkül.*
Comptes Rendus des Séances de la Société des Sciences et des Lettres de Varsovie, Classe III, 23(1930), 30–50.

[14] MAL'CEV A.: *Untersuchungen aus dem Gebiete der mathematischen Logik.*
Matiematičeskij Sbornik, 1(1936), 32–336.

[15] MAUTNER F. I.: *Logic as invariant theory, an extension of Klein's Erlangen Program.*
American Journal of Mathematics, 68(1946), 345–386.

[16] MCKINSEY J. C. C.: *Solution of the decision problem for Lewis systems S2 and S4 with an application to topology.*
Journal of Symbolic Logic, 6(1941), 117–134.

[17] MOSTOWSKI A.: *Proofs of non-deducibility in intuitionistic functional calculus.*
Journal of Symbolic Logic, 13 (1948), 204–207.

[18] RASIOWA H., SIKORSKI R.: *A proof of the completeness theorem of Gödel.*
Fundamenta Mathematicae, 37 (1950), 193–200.

[19] RASIOWA H., SIKORSKI R.: *The mathematics of metamathematics.*
Warszawa 1968.

[20] REICHBACH J.: *O pełności węższego rachunku funkcyjnego.*
Studia Logica, 2(1955), 213–228.
[21] RIEGER L.: *On free $\aleph_\xi$-complete Boolean algebras.*
Fundamenta Mathematicae, 38(1951), 35–52.
[22] RIEGER L.: *On countable generalized σ-algebras, with a new proof of Gödel's completeness theorem.*
Czechoslovak Mathematical Journal, 1(76) (1951), 29–40.
[23] STONE H. M.: *The theory of representations for Boolean algebras.*
Transactions of the American Mathematical Society, 40(1936), 37–111.
[24] SURMA S. J.: *Cztery studia z metamatematyki.*
Studia Logica, 23(1968), 80–114.
[25] SURMA S. J.: *Some results in metamathematical non-effectiveness.*
Universitas Iagellonica Acta Scientiarum Litterarumque, CCVIII, Schedae Logicae, 4(1969), 31–37.
[26] TARSKI A.: *Über einige fundamentale Begriffe der Metamathematik.*
Comptes Rendus des Séances de la Société des Sciences et des Lettres de Varsovie, Classe III, 23(1930), 22–29.
[27] TARSKI A.: *Une contribution à la théorie de la mesure.*
Fundamenta Mathematicae, 15(1930), 42–50.
[28] TARSKI A.: *Pojęcie prawdy w językach nauk dedukcyjnych.*
Travaux de la Société des Sciences et des Lettres de Varsovie, Classe III, 34(1933), 116.
[29] TARSKI A.: *Grundzüge des Systemenkalküls.* Erster Teil.
Fundamenta Mathematicae, 25(1935), 503–526.
[30] ULAM K.: *Concerning functions of sets.*
Ibid., 14(1929), 231–233.
[31] WAJSBERG M.: *Metalogische Beiträge.*
Wiadomości Matematyczne, 43(1937), 1–38.

ANDRZEJ WROŃSKI

ON THE OLD AND NEW METHODS OF INTERPRETING QUANTIFIERS [1]

The aim of this paper is to make a short review of the most important results dealing with the meaning of quantification. The quantifier expressions were used as operators binding variables in the works of Bolzano and Cauchy as soon as the beginning of XIXth century but using them Bolzano and Cauchy were guided only by logical intuition. First attempt to formalize those intuitions was undertaken in 1879 by Frege [4] [2]. It is generally accepted that the concept of quantifier was worked out independently by Peirce [31]. It was Peirce who introduced the symbols Σ, Π used up to now to denote quantifiers. He also introduced the name „quantifier" which is better than the name „Klammerzeichen" of Hilbert and Ackermann [21]. Revealing of the geometric sense of quantification is credited to Schröder [35] who noticed that logical operations on formulas with variables running over the set of real numbers correspond to the simple geometrical operations performed on graphs of these formulas. For instance if the graph of the formula $P(x_1, x_2, ..., x_n)$ is the subset of the Cartesian product $\mathfrak{R}_1 \times ... \times \mathfrak{R}_n$ composed of all n-tuples of real numbers satisfying $P(x_1, x_2, ..., x_n)$ and $1 \leqslant i \leqslant n$, then the projection and the inner projection of this graph paralell to axis $\mathfrak{R}_i$ give the graph of the formula $\sum_{x_i} P(x_1, ..., x_n)$ and $\prod_{x_i} P(x_1, ..., x_n)$, respectively. In 1931 Kuratowski and Tarski [25] established an interesting relation between the geometric analogues of logical operations of the classical predicate calculus and the so-called projective sets. By means of generalizations introduced in 1953 by Rasiowa and Mostowski [33] it is possible to apply the Schröder's

---

[1] This paper is an extension of a report presented on April 29, 1967 to the XIIIth Conference of the History of Logic organized in Cracow by the Section of Logic, Polish Academy of Sciences, together with the Department of Logic of the Jagiellonian University.

Its summary was published in Polish in Ruch Filozoficzny, 27(1969), 44–46.

[2] For number in the square brackets cf. p. 258.

ideas also to the field of non-classical logics. According to the generalized geometric interpretation the graph of the formula $P(x_1, \ldots, x_n)$ is a subset of the Cartesian product $\mathfrak{X}\times\mathfrak{R}_1\times\ldots\times\mathfrak{R}_n$ such that $\mathfrak{X}$ is a topological space. If $\mathfrak{X}$ is the one-element space, then the interpretation becomes adequate for the classical predicate calculus. In order to get the interpretation adequate for the intuitionistic predicate calculus it is necessary to assume that the topological space $\mathfrak{X}$ satisfies some more complicated conditions. Generalized geometric interpretation in the above sense is closely connected with Mostowski's idea to interpret the formulas of the predicate calculus as a suitable mappings, i.e., the so called Mostowski's functionals. In the paper [30] Mostowski's functional assigned to a formula is the mapping which associates an element of a complete Brouwerian algebra with every pair composed of an interpretation of the extra-logical symbols in a given fixed domain and of a valuation of the variables in this domain. The infinite Brouwerian join and meet corresponds to the existential and the universal quantification, respectively.

The wide-spreading of the so called Lindenbaum's method and the model-theoretic conceptual apparatus worked out in 1933 by Tarski [36] make possible the transfer of the investigations concerned with the logical calculi to the field of algebra [3]. So after the pioneer works Mautner [28], Mostowski [30], Henkin [15], [16], Rasiowa and Sikorski introduce in [34] the concept of the Lindenbaum algebra of the predicate calculus which in a very natural way reveals the algebraic sense of quantification. Let V denote the set of variables, T the set of terms and F the set of formulas of the predicate calculus. Let the symbols $\sim$, $\vee$, $\wedge$, Σ, Π denote negation, disjunction, conjunction, existential quantifier and universal quantifier, respectively. We say that the formulas β and γ are equivalent if both $\sim\beta\vee\gamma$ and $\sim\gamma\vee\beta$ are derivable. We denote by $|\beta|$ the set of all formulas equivalent to formula β and by $\beta x/t$ the formula resulting by the correct substitution of the term t for the variable x in formula β. Let $|F| = \{|\beta|\}_{\beta\in F}$ and $Q = \big\{\{|\beta x/t|\}_{t\in T}\big\}_{\beta\in F, x\in V}$. It is visible that for any given formula β and a variable x the set of equivalence classes of all formulas resulting by the correct substitution for x in β belongs to Q. By means of the above conventions the Lindenbaum algebra of the predicate calculus may be described as the structure $\langle |F|, -, \cup, \cap, \bigcup|Q, \bigcap|Q\rangle$ such that:

(L1) $\quad -|\beta| = |\sim\beta|$

(L2) $\quad |\beta|\cup|\gamma| = |\beta\vee\gamma|$

(L3) $\quad |\beta|\cap|\gamma| = |\beta\wedge\gamma|$

[3] This volume contains the article of J. S. Surma dealing with the history of the Lindenbaum's method.

whereas the operations $\bigcup, \bigcap$ corresponding to logical quantifiers have the set Q as a domain and are defined by the following conditions:

(L4) $\bigcup \{|\beta x/t|\}_{t \in T} = |\sum_x \beta|$

(L5) $\bigcap \{|\beta x/t|\}_{t \in T} = |\prod_x \beta|$

If the predicate calculus in question satisfies classical axioms, then corresponding Lindenbaum algebra may be proved to be embeddable in a complete Boolean algebra with additional operations $\bigcup$ (infinite Boolean join) and $\bigcap$ (infinite Boolean meet).

Still more general approach to the problem of algebraic sense of quantification was worked out by Tarski and Thompson [4] in 1952 and the other by Halmos [5] in 1954. Tarski and Thompson introduced the concept of the cylindric algebra, i.e., the structure $\langle U, I, -, \cap, \bigcup, E\rangle$ such that $\langle U, -, \cap\rangle$ is a Boolean algebra whereas $\bigcup: I \to U^U$ such that every $i, j \in I$ and $\beta, \gamma \in U$ the following axioms are satisfied:

(1) $\bigcup_i 0 = 0$

(2) $\beta \leqslant \bigcup_i \beta$

(3) $\bigcup_i (\beta \cap \bigcup_i \gamma) = \bigcup_i \beta \cap \bigcup_i \gamma$

(4) $\bigcup_i \bigcup_j \beta = \bigcup_j \bigcup_i \beta$

The operation $E: I \times I \to U$ is such that for every $i, j, k \in I$ and $\beta \in U$ the following axioms are satisfied:

(E1) $E(i, i) = 1$

(E2) $E(i, j) = \bigcup_k (E(i, k) \cap E(j, k))$ where $i, j \neq k$

(E3) $\bigcup_i (\beta \cap E(i, j)) \cap \bigcup_i (-\beta \cap E(i, j)) = 0$ where $i \neq j$

It is easy to see that the operations $\bigcup$ and E are algebraic analogues of the existential quantifier and the identity predicate, respectively.

The Halmos' polyadic algebra differs from the cylindric only by having instead of E the operation $S: I^I \to U^U$ which plays the rôle of algebraic analogue of substitution. Thus, a polyadic algebra is the structure $\langle U, I, -, \cap, \bigcup, S\rangle$ satisfying the axioms (1), ..., (4) and such that for every $f, g \in I^I$ and $\beta, \gamma \in U$ the following axioms are satisfied:

(S1) $S_f(-\beta) = -(S_f \beta)$

(S2) $S_f(\beta \cap \gamma) = (S_f \beta) \cap (S_f \gamma)$

(S3) $S_f \beta = \beta$, if $f(i) = i$ for every $i \in I$

[4] Cf. [38] and also [2], [5], [17], [18], [19], [20], [22], [23], [24], [29], [32], [37].

[5] Cf. [6] and also [1], [2], [3], [5], [7], ..., [14], [26], [27], [39], [40].

(S4) $S_f S_g \beta = S_{f;g} \beta$ where $f; g$ denotes superposition of f and g

(S5) $S_f \bigcup_i \beta = S_g \bigcup_i \beta$, if $f(j) = g(j)$ for every $j \neq i$

(S6) $\bigcup_i S_f \beta = S_f \bigcup_j \beta$, if $f(j) = i$

The axioms (S1), (S2), (S3) show that for every $f \in I^I$ the mapping S_f is a Boolean endomorphism: $U \to U$ and S_f is the identity endomorphism, if f: $I \to I$ is an identity mapping. The concept of functional cylindric (polyadic) algebra plays a very important rôle in the representation theory for cylindric (polyadic) algebras. Having non-empty sets Y, Z and a complete Boolean algebra $\langle B, -, \cap \rangle$ one may construct the functional polyadic algebra with the universe U consisting of all functions: $Z^Y \to B$. The following conditions define Boolean operations $\div$ and $\cap$ in the universe U:

For every β, γ: $Z^Y \to B$ and $v \in Z^Y$

(F1) $(\div \beta)(v) = -(\beta(v))$

(F2) $(\beta \cap \gamma)(v) = \beta(v) \cap \gamma(v)$

The following conditions define the operation S: $Y^Y \to U^U$ and the operation $\uplus$: $Y \to U^U$:

For every $f \in Y^Y$, $i \in Y$, β: $Z^Y \to B$ and $v \in Z^Y$

(F3) $(S_f \beta)(v) = \beta(\bar{v})$ where $\bar{v}(j) = v(f(j))$ for every $j \in Y$

(F4) $\uplus_i \beta = \bigcup \{S_f \beta : f(j) = j$ for every $j \neq i\}$

The symbol $\bigcup$ in (F4) denotes the infinite Boolean join which always exists in the complete Boolean algebra B. It is matter of routine to check that the structure $\langle U, Y, \div, \cap, \uplus, S \rangle$ is a polyadic algebra, i.e., it satisfies the axioms (1), ..., (4) and (S1), ..., (S6). The problem of the relationship between cylindric and polyadic algebras was solved in 1957 by Galler [5] who found the method of transforming one of them into another.

References

[1] Bass H.: *Finite monadic algebras.* Proceedings of the American Mathematical Society, 9(1958), 258–268.

[2] Copeland A. H. Sr.: *Note on cylindric algebras and polyadic algebras.* Michigan Mathematical Journal, 3(1955–1956), 155–157.

[3] Daigneault A., Monk D.: *Representation theory for polyadic algebras.* Fundamenta Mathematicae, 52(1963), 151–176.

[4] Frege G.: *Begriffsschrift, eine der mathematischen nachgebildete Formelsprache des reinen Denkens.* Halle, 1879.

[5] Galler B. A.: *Cylindric and polyadic algebras.*
Proceedings of the American Mathematical Society, 8(1957), 176–183.
[6] Halmos P. R.: *Polyadic Boolean algebras.*
Proceedings of the National Academy of Sciences of the USA, 40(1954), 296–301.
[7] Halmos P. R.: *Algebraic logic,* I.
Compositio Mathematica, 12(1955), 217–249.
[8] Halmos P. R.: *Algebraic logic,* II.
Fundamenta Mathematicae, 43(1956), 255–325.
[9] Halmos P. R.: *The basic concepts of algebraic logic.*
American Mathematical Monthly, 63 (1956), 363–387.
[10] Halmos P. R.: *Algebraic logic,* III.
Transactions of the American Mathematical Society, 83(1956), 430–470.
[11] Halmos P. R.: *Algebraic logic,* IV.
Transactions of the American Mathematical Society, 86(1957), 1–27.
[12] Halmos P. R.: *The representation of monadic Boolean algebras.*
Duke Mathematical Journal, 26(1959), 447–454.
[13] Halmos P. R.: *Free monadic algebras.*
Proceedings of the American Mathematical Society, 10(1959), 219–227.
[14] Halmos P. R.: *Algebraic logic.*
New York 1962.
[15] Henkin L.: *A proof of completeness for the first order functional calculus.*
Journal of Symbolic Logic, 14(1949), 159–166.
[16] Henkin L.: *An algebraic characterization of quantifiers.*
Fundamenta Mathematicae, 37(1950), 63–74.
[17] Henkin L.: *The representation theorem for cylindrical algebras.*
Mathematical Interpretation of Formal Systems.
Amsterdam 1955, 85–97.
[18] Henkin L.: *La structure algébrique des théories mathématiques.*
Paris, 1956.
[19] Henkin L., Tarski A.: *Cylindric algebras.*
Summaries of talks presented at the Summer Institute of Symbolic Logic, 3(1957), 332–340.
[20] Henkin L., Tarski A.: *Cylindric algebras.*
Proceedings of Symposia in Pure Mathematics, II (1961), Lattice theory, 83–113.
[21] Hilbert D., Ackermann W.: *Grundzüge der theoretischen Logik.*
Berlin 1928.
[22] Kasner M.: *Les algèbres cylindriques.*
Bulletin de la Société Mathématique de France, 68(1958), 315–319.
[23] Kotas J., Pieczkowski A.: *On a generalized cylindrical algebra and intuitionistic logic.*
Studia Logica, 18(1966), 73–80.
[24] Kotas J., Pieczkowski A.: *A cylindrical algebra based on the Boolean ring.*
Studia Logica, 21(1967), 71–78.
[25] Kuratowski C., Tarski A.: *Les opérations logiques et les ensembles projectifs.*
Fundamenta Mathematicae, 17(1931), 240–248.
[26] Leblanc L.: *Représentation des algèbres polyadiques pour anneau.*
Comptes Rendus Hebdomadaires des Séances de l'Académie des Sciences, 250(1960), 4092–4094.
[27] Leblanc L.: *Nonhomogenous polyadic algebras.*
Proceedings of the American Mathematical Society, 13(1962), 59–65.

[28] MAUTNER F. I.: *Logic as invariant theory, an extension of Klein's Erlangen program.* American Journal of Mathematics, 68(1946), 345–386.

[29] MONK D.: *On the representation theory for cylindric algebras.*

[30] MOSTOWSKI A.: *Proofs of non-deducibility in intuitionistic functional calculus.* Journal of Symbolic Logic, 13(1948), 204–207.
Pacific Journal of Mathematics, 11(1961), 1447–1457.

[31] PEIRCE C. S.: *On the algebra of logic.* American Journal of Mathematics, 7(1885).

[32] PIECZKOWSKI A.: *On the equivalence of the calculus of dependent variables and the cylindrical algebra without diagonal elements.* Bulletin de l'Académie Polonaise des Sciences, 12(1964), 143–146.

[33] RASIOWA H., MOSTOWSKI A.: *O geometrycznej interpretacji wyrażeń logicznych.* Studia Logica, 1(1953), 254–269.

[34] RASIOWA H., SIKORSKI R.: *A proof of the completeness theorem of Gödel.* Fundamenta Mathematicae, 37(1950), 193–200.

[35] SCHRÖDER E.: *Vorlesungen über die Algebra der Logik.* Vol. III. Leipzig 1895.

[36] TARSKI A.: *Pojęcie prawdy w językach nauk dedukcyjnych.* Travaux de la Société des Sciences et des Lettres de Varsovie, Classe III, 34(1933).

[37] TARSKI A.: *A representation theorem for cylindric algebras.* Bulletin of the American Mathematical Society, 58(1952), 65–66.

[38] TARSKI A., THOMPSON F. B.: *Some general properties of cylindric algebras.* Bulletin of the American Mathematical Society, 58(1952), 65.

[39] VARSAVSKY O.: *Quantifiers and equivalence relations.* Revista Matemática Guyana, 2(1958), 29–51.

[40] WRIGHT F. B.: *Ideals in a polyadic algebra.* Proceedings of the American Mathematical Society, 8(1957), 544–546.

WŁADYSŁAW SZCZĘCH

L. RIEGER'S LOGICAL ACHIEVEMENT [1]

Ladislav (Swante) Rieger (June 25, 1916 – February 13, 1963) was the first mathematician in Czechoslovakia, whose specialization was mathematical logic and the foundations of mathematics. Although he was born in Malmö (Sweden), from his early childhood he lived in Czechoslovakia (the mountains of Roztoki, Żylina, Prague). He studied mathematics at the Carol's University in Prague, where his father Ladislav Rieger was later a professor of philosophy. Unfortunately he did not manage to complete his studies because the Second World War broke out. During the war-time Rieger began to prepare his doctoral dissertation entitled: „On ordered and cyclically ordered groups", thanks to which in 1946 he gained the title of doctor of natural sciences. From the end of the war till 1958 he was working in Czechoslovak Higher Technical School and then in the Mathematical Institute of Czechoslovak Academy of Sciences in Prague. In 1951 he was made Privatdocent of Mathematics and in 1959 he was given the title of doctor of mathematical-physical sciences [2].

At the beginning of his activity in the field of logic Rieger was under the influence of the paper [1] because his articles [4], [5] and [11] give answers to some of the problems presented there. The papers [5] and [11] are less interesting but the paper [4] is a very interesting and important Rieger's achievement. In the article [5] there was given a construction of Boolean algebra, which has no proper automorphism and in [11] a construction of the countable closure algebra generated by one element. This algebra is not free. In the paper [4] two fundamental theorems are

---

[1] This article is a summary of the lecture delivered on April 26, 1969 at the XVth Conference of the History of Logic organized in Cracow by the Section of Logic, Polish Academy of Sciences, together with the Department of Logic of the Jagiellonian University.

[2] Biographic details are drawn from the article [2] where one can also find the full bibliography of L. Rieger's papers.

proved: the theorem on the existence of the free $\aleph_\xi$-complete Boolean algebras and Rieger's theorem on σ-isomorphic representation of the $\aleph_0$-complete Boolean algebra (in other words σ-algebra) by minimal σ-field of Borel subsets of generalized Cantor's discontinuum. The conclusions drawn from the second of the statements discussed above are: the Loomis' theorem, the representation theorem of Lindenbaum-Tarski algebras and the Gödel-Malcev completeness theorem. It is perhaps worth mentioning that L. Rieger stayed in Warsaw in 1950 and there he took part in A. Mostowski's seminar. It gave opportunity to exchange opinions, among others, with H. Rasiowa and R. Sikorski who were also interested, at this time, in the representation of logico-algebraic structures.

In the papers [6] and [9] he made use of a new notion of generalized Boolean σ-algebra, that means the σ-algebra in which only some countable sets of elements, so-called distinguished sequences, have a sum and product. In the paper [6] it was proved that the countable generalized σ-algebra with a countable set of distinguished sequences (such, e.g., is Lindenbaum's algebra based upon Hilbert-Ackermann formalism) is represented by σ-field of the sets and this fact was used for the proof of the Gödel's completeness theorem. But in the article [9] this notion was used for the proof of the Lindenbaum's theorem on complete super-systems and as a conclusion the Gödel's completeness theorem and the Skolem-Löwenheim theorem were obtained. There was also presented a comparison of different methods for proving the Gödel's and the Lindenbaum's theorems to which one should also add Rieger's method in which one may distinguish the following main stages:

(1) showing that an algebra of certain type (different in various Rieger's papers) may be σ-isomorphically represented by certain σ-field of sets,

(2) showing that the Lindenbaum algebra is just such an algebra,

(3) concluding that the Lindenbaum algebra may be represented just by this σ-field of sets,

(4) defining σ-homomorphism of the Lindenbaum algebra in a two-element Boolean algebra (completeness theorem).

In the article [8] Rieger achieves some interesting results about Suslin algebras, among others, an extension of the Loomis' theorem for uncountable algebras. In addition he shows how Suslin algebra may be used for a formalization of the second-order predicate logic. Finally in his earliest paper in logic [3] Rieger gives an algebraic characteristics of the sequential version of the Brouwerian propositional logic.

Papers [7], [13], [12] and [14] form a separate group, the first three of which are a partial reflection of the results fully presented in a posthumously published monograph [14] edited by M. Katetov. Its main result is an algebraic characteristics of the Lindenbaum algebras. As it is a paper written by a mathematician one may easily observe a multi--level course of considerations. Together with purely logical topics their modifications, attained by adding an established set of axioms (specific for a given mathematical theory) are discussed. For instance, apart from the relation of the logical consequence, denoted by $\Rightarrow$, another relation is discussed: the relation of relative consequence $\underset{A}{\Rightarrow}$, A being a set of axioms. In connection with the above is a neccessity of taking into consideration, apart from the Lindenbaum algebra L, also an algebra L_A obtained by means of the relative consequence. From the logical point of view it should be stressed that Rieger treats the consequence as a binary relation, in symbols, $\Rightarrow \subset S \times S$ (where S is the set of all propositional expresions) and not as a function, in symbols, $\Rightarrow: 2^S \to 2^S$ as it is generally done. The relation of consequence is given by means of the recursive definition, similarly as the set of all propositional expressions S. The deduction theorem, which says that $X \Rightarrow Y$ is valid, if and only if $Z \Rightarrow (X \to Y)$ is valid for every Z, allows us to observe easily that the set of logical theses (the proposition which is a consequence of any Z is treated as a thesis) equals to the set of theses of the classical logic. However, certain rules of deduction, e.g., the rule of detachment in a standard formulation, cannot be formulated at all although one may obtain certain schemes similar to them.

With the help of algebraic methods Rieger characterizes both logical syntax and semantics, but the chapter on logical semantics in [14] remained unfinished and was not completed until after Rieger's death. First of all, he shows that the notion of free Boolean algebra is sufficient for describing quantifierless logical syntax. On the other hand Rieger's own conceptions are the notions of the indexed and the substituted Boolean algebras which are necessary for the full characteristics of logical syntax with quantifiers. Two following theorems are basic:

(I) The Lindenbaum algebra L is i-isomorphic to a certain $\langle B, Z_I \rangle$ free indexed Boolean algebra with the set Z of free generators.

(II) The Lindenbaum algebra L_A is i-isomorphic to a certain $\langle B, Z_I \rangle$ indexed substituted Boolean algebra.

The following definitions explain the above used terms:

i-isomorphism is an isomorphism preserving countable products and sums.

$\langle B, Z_I \rangle$ is called the indexed Boolean algebra, if and only if

(1) B is a Boolean algebra

(2) Z_I is a system of indexations, i.e., mappings $\theta\colon I^n \to B$, I is

a set of indices, B is a set of the elements of the algebra, satisfying the conditions:

(a) Z_I is closed under sums, products and complementations of indexations,

(b) Z_I is closed under infinite sums and products performed upon the indices,

(c) for every indexation θ, if two of its indices are identified, then a new indexation is obtained,

(d) the sum of the counterdomains of all the indexations is equal to the universe of the algebra B.

$\langle B, Z_I\rangle$ is a free algebra with the set Z of free generators if and only if

(1) $Z \subset Z_I$

(2) Z generates Z_I

(3) each mapping $f\colon Z \to Z_I^*$, where $\langle B^*, Z_I^*\rangle$ is an indexed Boolean algebra, can be extended to an i-homomorphism $H_f\colon \langle B, Z_I\rangle \to \langle B^*, Z_I^*\rangle$ such that for every $\theta \in Z_I$

$$H_f\big(\theta(\xi_1, \ldots, \xi_n)\big) = f(\theta)(\xi_1, \ldots, \xi_n).$$

Finally, $\langle B, Z_I\rangle$ is called an indexed substituted Boolean algebra, if and only if Z_I satisfies the condition of substitution, i.e., if for every $\Phi, \Psi \in Z_I$, every $\sigma\colon I \to I$ and every $\varrho_1, \ldots, \varrho_n, \varepsilon_1, \ldots, \varepsilon_m \in I$:
if

$$\Phi(\varrho_1, \ldots, \varrho_n) = \Psi(\varepsilon_1, \ldots, \varepsilon_m),$$

then

$$\Phi\big(\sigma(\varrho_1), \ldots, \sigma(\varrho_n)\big) = \Psi\big(\sigma(\varepsilon_1), \ldots, \sigma(\varepsilon_m)\big).$$

The notions of the indexed and the substituted indexed Boolean algebras are similar to those of the cylindric algebras in the sense of Tarski-Thomson and the polyadic algebras in the sense of Halmos, though no closer relations among them are familiar to me.

In this part of the paper which is devoted to an algebraic characteristics of the logical semantics, Rieger's attention is focused on the problems of completeness of the first-order predicate calculus and of interpretability connected with construction of models. He formulates the completeness theorem in the following form: the product of all i-prime ideals, i.e., the prime ideals closed under infinite products of the Lindenbaum algebra is the unit set containing the unity of this algebra.

Then he introduces the notion of i-prime over-ideal of the Lindenbaum algebra and proves a theorem of its existence. It allows us to show that one can treat the theory of models as, according to Rieger's words, a theory of i-homomorphic representation of the indexed substituted

Lindenbaum algebras by the algebras of infinitely dimensional cylinders with finitely dimensional basis.

To sum up, it should be stated that Ladislav Rieger's achievement in the field of logic is not very large but very valuable. Rieger wrote various papers on some basic problems which are still interesting for the logicians. Rieger was one of those who helped to solve these problems, e.g., the problem of representation by Cantor's discontinuum, and achieved many valuable results.

References

[1] Birkhoff G.: *Lattice theory.* New York 1948.

[2] Čulik K.: *Žizn' i tworčestwo Ladislawa Riegera.* Czechoslovak Mathematical Journal, 14(89)(1964), 629–633.

[3] Rieger L.: *On the lattice theory of Brouwerian propositional logic.* Acta Facultatis Rerum Naturalium Universitatis Carolinae, 189(1949), 1-40.

[4] Rieger L.: *On free $\aleph_\xi$-complete Boolean algebras.* Fundamenta Mathematicae, 38(1951), 35–52.

[5] Rieger L.: *Some remarks on automorphisms in Boolean algebras.* Fundamenta Mathematicae, 38(1951), 209–216.

[6] Rieger L.: *O sčëtnych obobščënnych σ-algebrach i novom dokazatelstve teoremy Gëdelja o polnote.* Czechoslovak Mathematical Journal, 1(76)(1951),33–49.

[7] Rieger L.: *O algebře nižšiho predikátoveho počtu.* Praha 1951.

[8] Rieger L.: *Ob algebrach Suslina i ich predstavlenii.* Czechoslovak Mathematical Journal, 5(80)(1955), 99–142.

[9] Rieger L.: *O jedne základni vědě matematické logiky.* Časopis pro pěstováni matematiky, 80(1955), 217–231.

[10] Rieger L.: *O některých základnich otázkách matematické logiky.* Časopis pro pěstováni matematiky, 81(1956), 342–351.

[11] Rieger L.: *Zametka o t. naz. svobodnych algebrach s zamykanijami.* Czechoslovak Mathematical Journal, 7(82)(1957), 16–20.

[12] Rieger L.: *Matematická logika.* Praha 1961.

[13] Rieger L.: *Zu den Strukturen der klassischen Prädikatenlogik.* Zeitschrift für mathematische Logik und Grundlagen der Mathematik, 10(1964), 121–138.

[14] Rieger L.: *Algebraic methods of mathematical logic.* Prague 1967.

III. CONTRIBUTIONS OF THE HISTORY OF SOME SET-THEORETICAL AND METHODOLOGICAL TOPICS

JERZY PERZANOWSKI

THE DEVELOPMENT OF CANTOR'S DEFINITION OF THE SET [1]

A notion of „set" (or the notions such as the „class", „collection", „family", „system", „plurality", „multitude", „aggregate" or „complex" which are usually used interchangeably) is a fundamental notion in deductive sciences; it is also often used in other sciences, e.g., in law and social sciences. Though, *prima facie* clear and distinct this notion, however, when used carelessly may lead to the well-known set-theoretical antinomies. Hence came the great importance of the philosophical analysis of the notion „set".

The aim of this paper is to point out to the characteristic evolution of Cantor's idea of the set in which G. Cantor under the influence of the set-theoretical antinomies discovered by himself modified the definition of the set which he had originally accepted and, in addition, modified his ideas about the absolute character of the set theory [2].

In his basic paper entitled „Grundlagen einer allgemeinen Mannigfaltigkeitslehre" written in 1883 Cantor presented (cf. [3], p. 204) the following definition of the set: „[...] Unter einer «Mannigfaltigkeit» oder «Menge» verstehe ich nämlich allgemein jedes Viele, welches sich als Eines denken lässt, d. h. jeden Inbegriff bestimmter Elemente, welcher durch ein Gesetz zu einem Ganzen verbunden werden kann, und ich glaube hiermit etwas zu definieren, was verwandt ist mit dem Platonischen εἶδος oder ἰδέα [...]". The above quotation may be translated into English as follows: „[...] By «multitude» or «set» I mean every plurality which can be thought of as one (oneness), i.e., every integral approach to some

[1] This paper is an extended version of the lecture delivered by the author on April 23, 1966 to the XIIth Conference of the History of Logic organized in Cracow by the Section of Logic, Polish Academy of Sciences, together with the Department of Logic of the Jagiellonian University.

Its summary was published in Polish in Ruch Filozoficzny, 26(1968), 227–230.

[2] Obviously, I shall here discuss the sets understood distributively, as only this understanding of sets is the subject of the set-theoretical investigations.

elements which by law can be bound in unity, and I think that by the same I define something which is related to Platonic εἶδος or ἰδέα [...]". This definition was repeated with some modification by Cantor in the paper entitled „Beiträge zur Begründung der transfiniten Mengenlehre" written in 1895 (cf. [3], p. 282) where he says: „[...] Unter einer «Menge» verstehen wir jede Zusammenfassung *M* von bestimmten wohlunterschiedenen Objekten *m* unserer Anschauung oder unseres Denkens (welche die «Elemente» von *M* genannt werden) zu einem Ganzen [...]". The following translation of the passage may be given: „[...] By a «set» we shall understand any collection into a whole *M* of definite distinguishable objects *m* (which will be called «elements» of *M*) of our intuition or thought [...]".

It is obvious that this definition is closely connected with the traditional definition where sets are treated as the extensions of general notions (or names). This definition has two main faults, namely, it is too broad and not clear.

Let us analyze this definition. The set is a multitude of objects (which corresponds to what we usually mean by the word „plurality" derived from the adjective „plural"), i.e., a collection of objects whose number is not precisely given. We demand these subjects to be determined and well distinguished, i.e., to make it possible for us to decide whether an object is or is not an element of our set. The term „object" is here very widely understood as an object of our appearances. It does not result from this definition whether the sets, as the conceptualists would think, are the purely intentional entities (that seems to be suggested by the phrase „[...] can be thought [...]"); or it neither seems, however, to result whether the sets are beings of some mode of existence beyond-thought. The following phrase is also important „[...] which by law can be bound in unity [...]". Here, it is clear that Cantor was conscious of the fact that the elements of the set cannot be quite arbitrary but only the objects similar to one another in some respect can be elements of this set. This idea was developed by Cantor in the second version of his definition.

Not much is said in the definition about the nature of the elements in the set and about some specific properties of the set, e.g., about its structure. It is even unknown whether the sets, as presented by Cantor, are the only elements of the plurality while these elements are mentally combined together or whether these elements are combined somehow, in some order. But we known that Cantor treated sets as unique unities endowed with peculiar properties and combined of the peculiarly endowed elements. Upon thus endowed sets Cantor performed some abstractions (which were marked in the symbolism accepted by him). It must be remembered that he either abstracted from the nature of the elements obtaining the sets with some structure or he abstracted from the nature of the elements and the order in which the elements are given and then he obtained the sets only with the property of some cardinality. The sets in Cantor's approach cannot be reduced completely to their elements.

It should be stressed here that Cantor's definition does not allow us to accept the empty set as a set since, according to Cantor, the set is a multitude of elements whereas the set which has no elements is not a multitude. But we can introduce without any special difficulties the empty set into set theory in many different ways, e.g., as G. Frege did it, we may treat the empty set as an extension of the propositional function which cannot be fulfilled by any object or we can treat the empty set as a result of applying the operation of the product of two sets for disjoint sets.

It can also be observed that Cantor's definition is too broad, as it gives no rigorous conditions which the objects had to fulfill to be treated as a unity. Let us observe that Cantor's definition admits such sets as the set of all possible objects (determined, e.g., as a denotation of the name „something"), the set of all sets, the set of all ordinal numbers or the set of all cardinal numbers. No obstacles can be observed to take such sets into consideration. In particular, we have satisfactory criteria to include the elements to the above mentioned sets, e.g., to the set of all sets we include all what is a set, i.e., all what possesses the elements considered as the elements of a whole; to the set of all ordinal numbers we shall include every ordinal number etc.

But consideration of such sets leads us to the so-called set-theoretical antinomies which originate from the unsatisfactorily defined term „set".

Not later than in 1895 Cantor discovered the first antinomy in the discussed set-theoretical system. In 1896 he wrote to D. Hilbert about it (cf. [3], p. 470, and [2]). In 1897 the Italian mathematician C. Burali-Forti independently of Cantor discovered and published this antinomy. Today it is known as the Burali-Forti's antinomy. In the course of the next two years Cantor discovered all the other set-theoretical antinomies, with the exception of so-called Russell's antinomy. But it should be added that Cantor's proof of his well-known theorem about the power of the family of all subsets of any set made it possible for Russell to formulate this antinomy resulting from a simple modification of Cantor's reasoning.

The question of the modification of Cantor's original theory in order to avoid the antinomies became very urgent since in the nineties of the nineteenth century the set theory found numerous applications in various disciplines of the classical mathematics.

Cantor's first attempt to avoid the set-theoretical antinomies was carried on by means of redefinition of the notion of the set. This redefinition aimed at leaving out of the consideration the non-consistent sets.

In July 1899 (cf. [3], p. 443–444) in the letter to Dedekind Cantor gave the following definition of the set: „[...] Gehen wir von dem Begriff einer bestimmten Vielheit (eines Systems, eines Inbegriffs) von Dingen aus, so hat sich mir die Notwendigkeit herausgestellt, zweierlei Vielheiten (ich meine immer bestimmte Vielheiten) zu unterscheiden. Eine Vielheit kann nämlich so beschaffen sein, dass die Annahme eines «Zusammenseins» aller ihrer Elemente auf einen Widerspruch führt, so dass es unmöglich ist, die Vielheit als eine Einheit, als «ein fertiges Ding» aufzufassen. Solche Vielheiten nenne ich absolut unendliche oder inkonsistente Vielheiten. Wie man sich leicht überzeugt, ist z. B. der «Inbegriff alles Denkbaren» eine solche Vielheit; später werden sich noch andere Beispiele darbieten. Wenn hingegen die Gesamtheit der Elemente einer Vielheit ohne Widerspruch als «zusammenseiend» gedacht werden kann, so dass ihr Zusammengefasst werden zu «einem Ding» möglich ist, nenne ich sie eine konsistente Vielheit oder «Menge». (Im Französischen u. Italienischen wird dieser Begriff durch die Worte «ensemble» und «insieme» treffend zum Ausdruck gebracht.) Zwei äquivalente Vielheiten sind entweder beide «Mengen», oder beide inkonsistent. Jede Teilvielheit einer Menge ist eine Menge. Jede Menge von Mengen ist, wenn man die letzteren in ihre Elemente auflöst, auch eine Menge [...]". This might be translated as follows: „[...] Let us start with an idea of a plurality (a system, a whole of a thing). It seems necessary to distinguish between the two kinds of plurality (I mean a plurality of some kind). Namely, a plurality may be formed in such a way that to accept the coexistence of this elements meets a resistance so that it is impossible to apprehend such plurality as a unity (oneness), «a ready thing». I call this kind of pluralities the absolutely infinite or the non-consistent pluralities. As it is easy to notice just this plurality is, e.g., «a whole of this what is possible to think of», later on we shall give some other examples of this kind of plurality. But, on the other hand, when a whole of the elements in a plurality can be thought of as «conjunct» so that it is possible to bring them together as a whole, as «one thing» then I call it a consistent plurality or a «set». (In French and Italian this notion is accurately expressed by the word «ensemble» and «insieme».) Two equivalent pluralities are both either sets or they are both non-consistent. Every subsets of a set is a set itself. Every set of sets, when the latter ones are reduced to their elements, is also a set [...]".

The use of the word „plurality" in this context is explained by Cantor, in such a way he pays attention to „the pluralities of things not connected with each other", i.e., „such pluralities in which the removal of any element or more elements has no influence on the further existence of the remaining elements [...]". It can be noticed how well has Cantor known restrictions of the so-called non-predicative definition. The term „plurality" is here a broader term than the term „set". Then Cantor points out that, e.g., a plurality of all cardinal numbers or of all ordinal numbers or of all sets is a non-consistent plurality, i.e., it is not a set in the sense previously defined.

In the first version of his definition of the set, Cantor treated the set interchangeably with the notion of plurality (multitude) of some things. The appearance of the antinomy made Cantor accept distinction between two kinds of the pluralities: a non-consistent pluralities, that is, such ones whose elements cannot be united into one „without objection" and the so-called consistent pluralities, whose elements can be united in a whole and of these only the latter ones are now called by him sets.

There arises the question whether, having accepted the additional condition for the set to be „consistent", we do not carelessly leave out of the consideration those sets which should be the objects of our investigations. In particular, it is not clear whether the sets to which one assigns alephs as cardinal numbers are the sets in the sense previously exposed, i.e., whether they are consistent pluralities. Cantor points out (cf. [3], p. 447) that the same question concerns also the finite pluralities, for which a proof of their consistence cannot also be given, whereas for finite sets their consistence is obvious and undoubted. This thought is, according to Cantor, an axiom of the classical arithmetics and he thinks that an analogical axiom should be accepted for the transfinite arithmetics.

The axiom of consistence accepted here is undoubtedly the most fundamental axiom in the mathematics.

All other ways accepted in the set theory to avoid the antinomies have a similar postulative character. Introducting the postulates of this type into set theory deprives it of this absolute character which G. Cantor wanted to see in it and, moreover, it exposes to serious criticism on the basic for Cantor's („naive") set theory potential set.

Various ways were given later to avoid the set-theoretical antinomies. Almost all of them (except St. Leśniewski's mereology) had a tendency to avoid the antinomies by imposing either directly or by linguistic restrictions some confinements on the freedom of formation of the sets so as to eliminate, accepting these restrictions, the „antinomial" sets from the set theory.

The postulative character of the Cantor's way out of the problem of antinomies indicates that the mode of avoiding the antinomies initiated by E. Zermelo (cf. [11]), on the grounds of the axiomatized set theory is the best to conform with Cantor's intentions.

Of the different axiomatics of the set theory, the closest to Cantor's intentions are these in which — after John von Neumann — the distinction is made between two kinds of the set-theoretical entities; only upon the entities of one kind the majority of the standart set-theoretical operations is admitted to be performable. In the axiomatization due to P. Bernays and K. Gödel (cf. [1], [4], [10]), e.g., there are two kinds of the set theoretical entities characterized: the classes and the sets. At the same time the classes (contrary to the sets) can not be the elements of any other set theoretical objects; the majority of the set-theoretical operations is not performable upon them. In the axiomatics given by A. Morse (cf. [5], [8], [9]) a classifier is characterized, apart form the sets, which assigns to any set theoretical formula the totality of objects satisfying that formula; at the same time it sometimes happens that the entities given by this classifier are the sets and all the set-theoretical operations are the sets and all the set-theoretical operations are performable exclusively upon them (like at Bernays and Gödel).

In view of the fact that Cantor characterized the non-consistent pluralities as the entities upon which (contrary to the consistent entities, i.e., the sets) the standart set-theoretical operations are not performable, the conclusion is valid that Cantor, in his letter to Dedekind quoted above, anticipated the axiomatization of the set theory of the kind discussed above.

References

[1] Bernays P.: *A system of axiomatic set theory.* Journal of Symbolic Logic, (2)1937.

[2] Bernstein F.: *Über die Reihe der transfiniten Ordnungszahlen.* Mathematische Annalen, 60(1905).

[3] Cantor G.: *Gesammelte Abhandlungen.* Berlin 1932.

[4] Gödel K.: *The consistency of continuum hypothesis.* Princeton 1940.

[5] Kelley J. L.: *General topology.* Princeton 1955.

[6] Meschkowski H.: *Probleme des Unendlichen. Werk und Leben Georg Cantors.*

[7] Miedviediev F. A.: *Rozvitie teorii množestv v XIX viekie.* Moskwa 1965.

[8] Morse A.: *A theory of sets.* New York 1965.

[9] Mostowski A.: *Constructible sets with applications.* Warszawa 1969.

[10] Neumann von J.: *Die Axiomatisierung der Mengenlehre.* Mathematische Zeitschrift, 27(1928).

[11] Zermelo E.: *Untersuchungen über die Grundlagen der Mengenlehre.* Mathematische Annalen, 65(1908).

PIOTR KOSSOWSKI

ON THE ORIGINS OF THE SET-THEORETICAL CONCEPT OF RELATION[1]

As soon as set-theory, as a means of examining the foundations of mathematics was generally accepted, there arose a problem of the set-theoretical characterization of the concept of relation. In connection with the necessity of preserving the purity of types, the way of defining the notion of an ordered pair in the system of Russell and Whitehead does not obtrude itself. Thus, numerous informal explanations presented to us in the text of *Principia Mathematica*, let us guess what the authors mean but they cannot be at any rate recognized as correct definitions of the ordered pair. Such a definition was given later by Wiener in [5].

The question which Wiener considered can be summarized as follows: how should a set of a certain definite type, e.g., (γ) be chosen for two arbitrary types α and β in such a way that this set might have been adopted, by definition, as a set of the relational type (α, β)[2]. In other words, in the language of *Principia Mathematica*, it might be stated thus: how should a formula $\Phi^*(z^\gamma)$ with one free variable z^γ, such that

$$\bigwedge_{x^\alpha} \bigwedge_{y^\beta} [\Phi(x^\alpha, y^\beta) \equiv \Psi(x^\alpha, y^\beta)] \equiv \bigwedge_{z^\gamma} [\Phi^*(z^\varkappa) \equiv \Psi^*(z^\gamma)]$$

be chosen for an arbitrary formula $\Phi(x^\alpha, y^\beta)$ with two free variables x^α and y^β.

If it is possible to give a method of choosing the formula $\Phi^*(z^\gamma)$ for every formula $\Phi(x^\alpha, y^\beta)$ then its extension, i.e., $\hat{z}^\gamma\Phi^*(z^\gamma)$ will be a set

[1] It is a summary of report presented by the author on April 25, 1970 to the XVIth Conference of the History of Logic organized in Cracow by the Section of Logic, Polish Academy of Sciences, together with the Department of Logic of the Jagiellonian University.

[2] The set in question must fulfill the principle of purity of types.

of the type (γ) to which, by definition, the relational type (α, β) may be assigned by means of the above conditions.

The above task is accomplished by Wiener in the following way. In case of the formula $\Phi(x^\alpha, y^\alpha)$ (in which both variables are of the same type) he puts:

$$\Phi^*(z^\gamma) = \bigvee_{x^\alpha} \bigvee_{y^\alpha} \left[\Phi(x^\alpha, y^\alpha) \wedge z^\gamma = \left\{\{\{x^\alpha\}\} \cup \{\emptyset\}\right\} \cup \left\{\{\{y^\alpha\}\}\right\}\right],$$

obviously

$$z^\gamma = \left\{\{\{x^\alpha\}, \emptyset\}, \{\{y^\alpha\}\}\right\},$$

then $\gamma = \big(((\alpha))\big)$.

It may be asked why Wiener did not put simply[3]:

$$z^\gamma = \left\{\{x^\alpha\} \cup \{\emptyset\}\right\} \cup \left\{\{y^\alpha\}\right\} = \left\{\{x^\alpha, \emptyset\}, \{y^\alpha\}\right\}.$$

It would not be proper because in case $\alpha = 0$ the set $\{x^\alpha\} \cup \{\emptyset\}$ does not exist as there is no empty set of type 0. In a similar way the necessity of the assumption for the two variables to be of the same type may be explained. If $\alpha \neq \beta$, then the sets $\left\{\{\{x^\alpha\}, \emptyset\}\right\}$ and $\left\{\{\{y^\beta\}\}\right\}$ are also of different types and they cannot be summed up. In this case typical complications force Wiener to give separate definitions for every two different types. And so, for $\Phi(z^{(\alpha)}, y^\alpha)$ Wiener puts:

$$\Phi^*(z^\gamma) = \bigvee_{x^{(\alpha)}} \bigvee_{y^\alpha} \left[\Phi(x^{(\alpha)}, y^\alpha) \wedge z^\gamma = \left\{\{\{x^{(\alpha)}\}, \emptyset\}, \left\{\{\{y^\alpha\}, \emptyset\}\right\}\right\}\right],$$

but for $\Phi(x^{((\alpha))}, y^\alpha)$:

$$\Phi^*(z^\gamma) = \bigvee_{x^{((\alpha))}} \bigvee_{y^\alpha} \left[\Phi(x^{((\alpha))}, y^\alpha) \wedge z^\gamma = \left\{\{\{x^{((\alpha))}\}, \emptyset\}, \left\{\left\{\{\{y^\alpha\}, \emptyset\}, \emptyset\right\}\right\}\right\}\right],$$

and so on.

For $\Phi(x^\alpha, y^{(\alpha)})$:

$$\Phi^*(z^\gamma) = \bigvee_{x^\alpha} \bigvee_{y^{(\alpha)}} \left[\Phi(x^\alpha, y^{(\alpha)}) \wedge z^\gamma = \left\{\left\{\{\{x^\alpha\}, \emptyset\}, \emptyset\right\}, \{\{y^{(\alpha)}\}\}\right\}\right],$$

and for $\Phi(x^\alpha, y^{((\alpha))})$:

$$\Phi^*(z^\gamma) = \bigvee_{x^\alpha} \bigvee_{y^{((\alpha))}} \left[\Phi(x^\alpha, y^{((\alpha))}) \wedge z^\gamma = \left\{\left\{\left\{\{\{x^\alpha\}, \emptyset\}, \emptyset\right\}, \emptyset\right\}, \{\{y^{((\alpha))}\}\}\right\}\right].$$

One can ask why Wiener did not use simpler definitions. For example, for the formula $\Phi(x^{(\alpha)}, y^\alpha)$ one might put:

$$\Phi^*(z^\gamma) = \bigvee_{x^{(\alpha)}} \bigvee_{y^\alpha} \left[\Phi(x^{(\alpha)}, y^\alpha) \wedge z^\gamma = \left\{\{\{x^{(\alpha)}\}, \emptyset\}, \left\{\{\{y^\alpha\}\}\right\}\right\}\right].$$

---

[3] The notation used in the article does not visualize the typical ambiguity of the empty set $\emptyset$.

But such a definition would not preserve the condition: two identical relations have their domains and counter-domains of the same type. It may be illustrated by the following example:

Consider the formulas

$$\Phi(x^{(a)}, y^{a}) \quad \text{and} \quad \Phi'(x^{(a)}, z^{(a)}) = \bigvee_{y^{a}} [\Phi(x^{(a)}, y^{a}) \wedge z^{(a)} = \{y^{a}\}] .$$

In case of the simplified definition we arrive at the following equality:

$$\bigwedge_{x^{(a)}} \bigwedge_{z^{(a)}} \Phi'(x^{(a)}, z^{(a)}) = \bigwedge_{x^{(a)}} \bigwedge_{y^{a}} \Phi(x^{(a)}, y^{a}) ,$$

but the counter-domain of the relations corresponding to the formulas: $\Phi'(x^{(a)}, z^{(a)})$ and $\Phi(x^{(a)}, y^{a})$ have different types. Complications which Wiener had to face disappear in the system of Zermelo and Fraenkel. Therefore the Kuratowski's definition of the ordered pair which is suitable for this system is much simpler (cf. [3]).

Kuratowski's definition is yielded by general considerations of the concept of ordered family given by Hessenberg in [2]. According to Hessenberg a family P introduces an order in the set Z, if and only if

(1) $P \subset 2^{Z}$

(2) $\bigwedge_{X, Y \in P} (X \subset Y) \vee (Y \subset X)$

(3) $\bigwedge_{\substack{x, y \in Z \\ x \neq y}} \bigvee_{V \in P} (x \notin V \wedge y \in V) \vee (x \in V \wedge y \notin V)$

(4) $\bigwedge_{R \subset P} \bigcup R \in P$

(5) $\bigwedge_{R \subset P} \bigcap R \in P$

For a given family P fulfilling (1)–(5) the order $\underset{P}{<}$ in the set Z is defined by Hessenberg as follows:

$$x \underset{P}{<} y, \quad \text{if and only if} \quad \bigvee_{V \in P} (x \notin V \wedge y \in V)$$

To avoid some inaccuracies noticed in Hessenberg's definition of ordering family, Kuratowski proposes his own definition[4]:

The family $P \subset 2^{Z}$ introduces an order in the set Z, if and only if it is a maximal one fulfilling the condition (2) of the Hessenberg's definition.

By means of above definition Kuratowski proves that among the orders in the set Z and the ordering families there is a one to one cor-

[4] The discuss inaccuracies mentioned is beyond the subject of our paper.

respondence. Moreover, it is easy to observe that $M_1 = \{\{a, b\}, \{a\}\}$ and $M_2 = \{\{a, b\}, \{b\}\}$ are the only ordering families in the sense of Kuratowski in the two element set $Z = \{a, b\}$.

It seems natural to Kuratowski to adopt the following definition:

Family M_1 is an ordered pair in which a is the first element and b is the second. Simplicity and other advantages of Kuratowski's definitions may be at once noticed while compared with a similar one given by Hausdorff in [1]:

If elements (1) and (2) are different and none of them is identical either with a or with b, then the families $\{\{a, 1\}, \{b, 2\}\}$ and $\{\{a, 2\}, \{b, 1\}\}$ are ordered pairs.

References

1] HAUSDORFF F.: *Grundzüge der Mengenlehre.* Leipzig 1914.

[2] HESSENBERG G.: *Grundbegriffe der Mengenlehre.* Abhandlungen der Fries'schen Schule I, 4(1906), 674–685, Göttingen.

[3] KURATOWSKI C.: *Sur la notion de l'ordre dans la Théorie des Ensembles.* Fundamenta Mathematicae, 2(1921), 161–171.

[4] RUSSELL B., WHITEHEAD A. N.: *Principia Mathematica.* Vol I. Cambridge, England, 1910.

[5] WIENER N.: *A simplification of logic of relations.* Proceedings of the Cambridge Philosophical Society, 17(1914), 387–390.

STANISŁAW J. SURMA

A SURVEY OF VARIOUS CONCEPTS OF COMPLETENESS OF THE DEDUCTIVE THEORIES [1]

Considering that many metamathematical theorems are in fact theorems about different kinds of completeness of the deductive theories, I shall discuss in the present paper a few of such concepts of completeness which had been to a greater or lesser extent applied in metamathematics. This discussion seems to lead up to the conclusion about the incoherent and heterogeneous intuitions associated with the concept of completeness. At first we shall present two syntactical and purely combinatorial concepts of completeness and afterwards we are going to present the semantic, i.e., model-theoretic formulation of this concept. Using the analogy to the well-known concept of categoricity in power (cf. [10] and [23]) we weaken, in turn, the semantic versions of completeness to obtain the concept of the completeness in power. Two among them, completeness in power in the Robinson's sense, and completeness in power in the Tarski's sense, have permitted, i.a., to strengthen the Łoś-Vaught's theorem about categoricity in power. Another version of completeness in power, that in the Gödel's sense, leads up to separate theories containing as theorems only the true propositions in each model of the fixed power.

We apply the well-known set-theoretical symbols. By ϵ we denote the membership relation while by $\notin$ we denote the complement of this relation. By $=$, $\neq$, $\subset$ and $\subsetneq$ we denote the relations of equality, inequality, inclusion and proper inclusion between sets, respectively. The symbol $\emptyset$ denotes the empty set. The symbols $\bigcup$ and $\bigcap$ denote the sum and product of sets, respectively. By $\{A\}$ we denote the set whose

---

[1] This paper is an extension of the talk presented on April 23, 1966 to the XIIth Conference of the History of Logic organized in Cracow by the Section of Logic, Polish Academy of Sciences, together with the Department of Logic of the Jagiellonian University.

Its summary was published in Polish in Ruch Filozoficzny, 26(1968), 230–233.

only element is A. By the symbol $\aleph_0$ we denote the power (number of elements) of any denumerably infinite set.

We shall found our considerations below on the first order predicate calculus with identity and constant terms which from now on we shall simply call the first order predicate calculus. It may be described as follows.

The set $\{x_i\}_{i\in I_0}$, where $\overline{\overline{I_0}} \geqslant \aleph_0$, consists of the individual variable symbols. The set $\{c_i\}_{i\in I_1}$, where $\overline{\overline{I_1}} \geqslant 0$, consists of the individual constant symbols. The elements of the set $\{x_i\}_{i\in I_0} \cup \{c_i\}_{i\in I_1}$ are called terms. The set $\{P_i\}_{i\in I_2}$, where $\overline{\overline{I_2}} > 0$, consists of the predicate symbols. Using terms and predicate symbols one can build predicates of the form $P_i(t_1, t_2, \ldots, t_{n_i})$, where $t_1, t_2, \ldots, t_{n_i}$ are terms. We assume that P_i is the symbol of n_i-ary predicate. Among the predicate symbols we reckon the binary symbol of identity $=$ by means of which one can build predicates of the form $t_i = t_j$, where t_i, t_j are terms.

In case $\overline{\overline{I_0 \cup I_1 \cup I_2}} \leqslant \aleph_0$ we shall call the alphabet of the first order predicate calculus the denumerable alphabet. In the opposite case we shall call it the non-denumerable alphabet.

The smallest set composed of predicates and closed under logical constants $\neg$ (a one-argument symbol of negation), $\rightarrow$ (a two-argument symbol of implication), and $\bigwedge x$ (symbol of the universal quantifier binding the individual variable x) is called the set of the propositional formulas. We have also adopted certain convenient abbreviations. Namely, $A \vee B$ is an abbreviation of the formula $\neg A \rightarrow B$, formula $A \wedge B$ is an abbreviation of the formula $\neg(\neg A \vee \neg B)$ and finally the formula $\bigvee x A$ is an abbreviation of the formula $\neg \bigwedge x \neg A$. We shall denote by S the set of all sentences, i.e., of the propositional formulas which do not contain any free individual variables.

We shall denote by T any arbitrary set of constant terms. When $X \subset S$, the notation $T(X)$ will denote the set of all constant terms occurring in sentences belonging to set the X.

Within the set S we distinguish in one of the known ways, e.g., like in [6], a suitable number of propositions called axioms and we define the set of all theses of the first order predicate calculus as the smallest set containing these axioms and closed under the well-known rule of detachment. When $X \subset S$, then by $Cn(X)$ we denote the smallest set of sentences containing, beside the theses of the first order predicate calculus, the set X and closed under the rule of detachment, and we call it the set of consequences of set the X. The properties of the operation Cn are well-known (cf. [17], [18]). With the concept Cn (and the concepts of the general set theory) one can define a number of metamathematical concepts (cf. on the subject [17], [18]). Namely, let $X \subset S$. The set X

is said to be deductive system, in symbols, $X \in \mathfrak{S}$, if and only if X is closed under Cn, i.e., $Cn(X) \subset X$. The set X is said to be consistent, in symbols, $X \in \mathfrak{W}$, if and only if there exists no $A \in S$ such that $A \in Cn(X)$ and $\rightharpoondown A \in Cn(X)$. The set X is said to be deductively complete, in symbols, $X \in \mathfrak{V}$, if and only if for any $A \in S$ at least one of the two holds: $A \in Cn(X)$ or $\rightharpoondown A \in Cn(X)$.

Deductive theories which can be formalized by means of the first order predicate calculus are called elementary theories. In the opposite case they are called non-elementary.

Let U be an arbitrary non-empty set about which we assume that it constitutes the domain of ranging of the individual variables from the set $\{x_i\}_{i \in I_0}$. Let $\mathfrak{z}$ be a function (the so-called denotation or interpretation function) defined on the set $\{c_i\}_{i \in I_1} \cup \{P_i\}_{i \in I_2}$ and such that

(i) $\mathfrak{z}(c_i) = s_i$, where $s_i \in U$ and $i \in I_1$,

(ii) $\mathfrak{z}(P_i) = r_i$, where $r_i \subset U^{n_i}$ and $i \in I_2$.

The sequence

$$\mathfrak{M} = \langle U, \{s_i\}_{i \in I_1}, \{r_i\}_{i \in I_2} \rangle$$

is called a model of the described alphabet of the first order predicate calculus.

By the power of a model $\mathfrak{M}$ we mean the power of its universal set U.

By the symbol $E(\mathfrak{M})$ we denote the content of the model $\mathfrak{M}$, i.e., the set of all the sentences of S which are true in $\mathfrak{M}$ (for an exact definition of the truth cf. [19]). Let us notice that $E(\mathfrak{M}) \in \mathfrak{S} \cap \mathfrak{W} \cap \mathfrak{V}$. The expression $X \subset E(\mathfrak{M})$ means that the set X is true in the model $\mathfrak{M}$ or, in other words, the model $\mathfrak{M}$ verifies the set X.

According to the definition of the model $\mathfrak{M}$ the relation $\mathfrak{z}(=)$ is in $\mathfrak{M}$ a congruence, i.e., a reflexive, symmetric and transitive relation consistent with all the relations in $\mathfrak{M}$. Evidently, identity is a congruence relation but not conversely. A model $\mathfrak{M}$ is said to be normal provided that $\mathfrak{z}(=)$ is identity relation in it. In the sequel we shall restrict ourselves exclusively to normal models.

Models $\mathfrak{M} = \langle U, \{s_i\}_{i \in I_1}, \{r_i\}_{i \in I_2} \rangle$ and $\mathfrak{M}' = \langle U', \{s'_i\}_{i \in I_1}, \{r'_i\}_{i \in I_2} \rangle$ are said to be isomorphic, in symbols, $\mathfrak{M} iz \mathfrak{M}'$, if and only if there exists at least one one-to-one mapping φ of the set U onto the set U' and such that

(i) $\varphi(s_i) = s'_i$,

(ii) $\langle u_1, u_2, \ldots, u_{n_i} \rangle \in r_i$ if and only if $\langle \varphi(u_1), \varphi(u_2), \ldots, \varphi(u_{n_i}) \rangle \in r'_i$ for every $u_1, u_2, \ldots, u_{n_i} \in U$.

Let n be any positive integer number. Let us denote by σ_n and δ_n the sentence stating that there exist at least n different objects and the sentence stating that there exist exactly n different objects, respectively. The sentences σ_n and δ_n can be easily defined by means of the first order predicate calculus. The definitions may be as follows:

$$\sigma_n = \bigvee x_1 \bigvee x_2 \ldots \bigvee x_n \big(\neg(x_1 = x_2) \wedge \neg(x_1 = x_3) \wedge \ldots \wedge \neg(x_1 = x_n) \wedge \\ \wedge \neg(x_2 = x_3) \wedge \ldots \wedge \neg(x_{n-1} = x_n)\big)$$

$$\delta_n = \sigma_n \wedge \bigvee x_1 \bigvee x_2 \ldots \bigvee x_{n+1} \big((x_1 = x_2) \vee (x_1 = x_3) \vee \ldots \vee (x_1 = x_{n+1}) \vee \\ \vee (x_2 = x_3) \vee \ldots \vee (x_2 = x_{n+1}) \vee \ldots \vee (x_n = x_{n+1})\big).$$

It is easy to see that $\sigma_n \in E(\mathfrak{M})$ if and only if $\overline{\overline{\mathfrak{M}}} \geqslant n$, $\delta_n \in E(\mathfrak{M})$ if and only if $\overline{\overline{\mathfrak{M}}} = n$.

In the sequel we shall use the following well-known theorems.

THEOREM ON ISOMORPHIC MODELS (cf., e.g., [12]). *If* $\mathfrak{M}$ *iz* $\mathfrak{N}$, *then* $E(\mathfrak{M}) = E(\mathfrak{N})$.

GÖDEL-MALCEV'S COMPLETENESS THEOREM (cf. [5] and [11]). *If* $X \in \mathfrak{W}$, *then there exists at least one model* $\mathfrak{M}$ *such that* $X \subset E(\mathfrak{M})$.

„UPWARD" LÖWENHEIM'S THEOREM (cf. [1]). *If* $\overline{\overline{T(X)}} \leqslant \aleph_0$ *and if there exists at least one infinite model* $\mathfrak{M}$ *such that* $X \subset E(\mathfrak{M})$, *then for any* $\mathfrak{n} \geqslant \aleph_0$ *there exists at least one model* $\mathfrak{N}$ *of the power* $\mathfrak{n}$ *such that* $X \subset E(\mathfrak{N})$.

THEOREM (cf. [16]). *If* $\overline{\overline{T(X)}} \leqslant \aleph_0$ *and* $X \in \mathfrak{W}$ *and if* $\sigma_n \in Cn(X)$ *for any positive integer* n, *then for any* $\mathfrak{m} \geqslant \aleph_0$ *there exists at least one model* $\mathfrak{M}$ *of the power* $\mathfrak{m}$ *such that* $X \subset E(\mathfrak{M})$.

1. The most widely conceived notions of the completeness

The most widely conceived notions of the completeness of a deductive theory would, no doubt, consist in the fact that this theory would contain as its theorems all sentences which can be formulated in the language of this theory. Completeness would then be, however, a virtually useless concept. Below I am going to present some such conceptions of the completeness of which practical use has been done in the theory of deductive systems. Only one of these conceptions should be originated from the XIX century. All the rest belongs to the twentieth century.

2. Completeness conceived of as the deductive completeness in the sense of $\mathfrak{B}$

This concept of the completeness probably reaches back to Aristotelean times and is in common use (cf. [17], [18], [19] and [21]). The examples of the complete systems in this sense are the elementary theory of Boolean algebras without atoms, the elementary theory of the dense order without minimum and maximum and a number of other theories. This property has also the set of sentences $E(\mathfrak{M})$ for any $\mathfrak{M}$. The example of deductively incomplete systems are, i.a., the first order predicate calculus and the elementary theory of order, where the question of minimum or maximum has not been axiomatically resolved.

3. Maximal consistency or completeness in the Post's sense

This concept was introduced by E. Post in [14] in 1921. In the thirties it was adopted by Tarski in his consequence theory (cf. [17], [18], [19] and [21]). The system X is complete in Post's sense, symbolically, $X \in \boldsymbol{P}$, if and only if $X \cup \{A\} \notin \mathfrak{W}$ for an arbitrary $A \notin Cn(X)$.

Restricting oneself to the set of sentences, i.e., to a set of propositional formulas without free variables, it is easy to see that $\boldsymbol{P} = \mathfrak{V}$. However, on more extensive grounds, composed of all propositional formulas which can be constructed in the alphabet of the first order predicate calculus we have $\mathfrak{V} \not\subset \boldsymbol{P}$, for, e.g., the set of all propositional formulas true in the model $\mathfrak{M}$ belongs to $\boldsymbol{P}$, but not to $\mathfrak{V}$.

4. Categoricity or completeness in the Veblen's sense

System X is categorical, in symbols, $X \in \boldsymbol{V}$, if and only if, for any models $\mathfrak{M}$ and $\mathfrak{N}$ if $X \subset E(\mathfrak{M}) \cap E(\mathfrak{N})$, then $\mathfrak{M}$ *iz* $\mathfrak{N}$. The first order predicate calculus is obviously a non-categorical one. Non-categorical is also the theory of groups and many others. The examples of categorical theories are, i.a., affine geometry and Peano's arithmetics of natural numbers. These are, however, non-elementary theories. The so-called elementary deductively complete arithmetics of natural numbers, i.e., the set of sentences $E(\langle N, 0, 1, seq, +, \cdot, < \rangle)$ is non-categorical, where N is the set of all the natural numbers, 0 and 1 are the first and the second natural numbers, seq, $+$, $\cdot$ are the well-known operations of immediate successor, addition and multiplication in N while $<$ is the well-known ordering relation in N.

The concept of categoricity was introduced in 1904 by O. Veblen while carrying on the research concerned with the principles of geometry (cf. [24]). Also E. Huntington used this concept (cf. [3]). It should be noted that in mathematical publications one can find previous results stating the categoricity of different theories. Thus, e.g., R. Dedekind in 1888 in [4] cited the principal elements of the ordinary proof of categoricity of axioms (or, in Dedekind's own words, „Bedingungen") of the theory now known as Peano's arithmetics of natural numbers. The result proved by G. Cantor in 1895 in [2] virtually consists in the fact that a system of axioms („Merkmale" in Cantor's terminology) which forms his well-known definition of continuum, is a categorical one.

In the class of elementary deductive systems we have $V \not\subseteq \mathfrak{B}$. On the other hand, in the class of all deductive systems the concepts V and $\mathfrak{B}$ overlap, since, e.g., Peano's arithmetics of natural numbers belongs to V, but not, as we know, to $\mathfrak{B}$.

In close connection with the categoricity is the strict (or unique) categoricity and monotransformability introduced by Tarski in 1936 (cf. [20]). The system X is monotransformable, if and only if for any two models verifying X there is exactly one function which establishes the isomorphism of these models. For instance, the property of monotransformability is possessed by Peano's arithmetics of natural numbers. The system X is strictly (or uniquely) categorical, if and only if for any two models verifying X there is at most one function establishing the isomorphism of these models.

The concept of categoricity is of little use in research on elementary systems. This is due to the fact that no elementary deductive system verifiable by infinite models is categorical.

A system is categorical in power $\mathfrak{m}$, symbolically, $X \in V_{\mathfrak{m}}$, if and only if for any two models $\mathfrak{M}$ and $\mathfrak{N}$, if $\overline{\overline{\mathfrak{M}}} = \overline{\overline{\mathfrak{N}}} = \mathfrak{m}$ and if $X \subset E(\mathfrak{M}) \cap E(\mathfrak{N})$, then $\mathfrak{M}$ *iz* $\mathfrak{N}$. The concept of $V_{\mathfrak{m}}$ is weaker than V. It was introduced in 1954 independently by two researchers, J. Łoś in [10] and R. Vaught in [23]. The concept is highly relevant in the research on the elementary systems. This is testified to by the following:

THEOREM 1 (cf. [10] and [23]). *Let* $\overline{\overline{T(X)}} \leqslant \aleph_0$. *Then, if the elementary system* X *is verifiable only by infinite models and, moreover,* X *is categorical in some infinite power* $\mathfrak{m}$, *then* X *is deductively complete.*

The examples of categorical systems in denumerable power are, i. a., the elementary theory of dense order without minimum and maximum and the elementary theory of Boolean algebras without atoms. An example of the system non-categorical in any power, including denumerable power, is the above mentioned elementary deductively complete arithmetics of natural numbers.

The concepts above are not the only kinds of categoricity. A number of kinds of categoricity was examined by A. Grzegorczyk in [7] (cf. also [13]).

5. Completeness in the Tarski's sense

A system X is complete in the Tarski's sense, in symbols, $X \in \mathbf{T}$, if and only if for any models $\mathfrak{M}$ and $\mathfrak{N}$ if $X \subset E(\mathfrak{M} \cap E(\mathfrak{N})$, then

$E(\mathfrak{M}) = E(\mathfrak{N})$. This concept is a modification of the following concept of indiscernibility introduced by Tarski in [22] in 1950: models $\mathfrak{M}$ and $\mathfrak{N}$ are indiscernible, if and only if they have identical content, i.e., $E(\mathfrak{M}) = E(\mathfrak{N})$. For instance, the set $E(\mathfrak{M})$ itself which is non-categorical for any infinite $\mathfrak{M}$ has this property.

THEOREM 2. $\mathbf{T} = \mathfrak{V}$.

Proof. Let us assume that the system X is verified by two models having different contents. Because of the consistency of the contents of any models the set X has to be deductively incomplete, i.e., $X \notin \mathfrak{V}$. Now let us assume that $X \notin \mathfrak{V}$. Then there exists $A \in S$ such that both sets $X \cup \{A\}$ and $X \cup \{\rightarrow A\}$ are consistent. From the Gödel-Malcev's completeness theorem follows that there exist models having different contents which verify the set X, i.e., $X \notin \mathbf{T}$.

Using the analogy to the concept $V_{\mathfrak{m}}$, the concept of $\mathbf{T}$ can be weakened as follows: $X \in \mathbf{T}_{\mathfrak{m}}$, if and only if for any models $\mathfrak{M}$ and $\mathfrak{N}$ if $\overline{\overline{\mathfrak{M}}} = \overline{\overline{\mathfrak{N}}} = \mathfrak{m}$ and if $X \subset E(\mathfrak{M}) \cap E(\mathfrak{N})$, then $E(\mathfrak{M}) = E(\mathfrak{N})$. Applying the concept $\mathbf{T}_{\mathfrak{m}}$ one can strengthen the theorem 1 as follows:

THEOREM 3. *Let $\overline{\overline{T(X)}} \leqslant \aleph_0$. Then if the elementary system X is verifiable only by infinite models and, moreover, there exists $\mathfrak{m} \geqslant \aleph_0$ such that $X \in \mathbf{T}_{\mathfrak{m}}$, then X is deductively complete.*

Proof. Let us assume that $X \notin \mathfrak{V}$. Then there exists $A \in S$ such that $A \notin X$ and $\rightarrow A \notin X$. Hence both sets $X \cup \{A\}$ and $X \cup \{\rightarrow A\}$ are consistent. According to Gödel-Malcev's completeness theorem there exist models $\mathfrak{M}$ and $\mathfrak{N}$ such that $X \cup \{A\} \subset E(\mathfrak{M})$ and $X \cup \{\rightarrow A\} \subset E(\mathfrak{N})$. In view, of the assumption of the theorem 3 the system X is verifiable only by infinite models. Hence $\mathfrak{M}$ and $\mathfrak{N}$ are infinite. Hence according to the „upward" Löwenheim's theorem for any infinite power $\mathfrak{m}$ there exist models $\mathfrak{M}'$ and $\mathfrak{N}'$ of the power $\mathfrak{m}$ such that $X \cup \{A\} \subset E(\mathfrak{M}')$ and $X \cup \{\rightarrow A\} \subset E(\mathfrak{N}')$. In view of the assumption of the theorem 3 one has $X \in \mathbf{T}_{\mathfrak{m}}$ for some $\mathfrak{m} \geqslant \aleph_0$. Hence $E(\mathfrak{M}') = E(\mathfrak{N}')$. But on the other hand $A \in E(\mathfrak{M}')$ and $\rightarrow A \in E(\mathfrak{N}')$, i.e., $E(\mathfrak{M}') \neq E(\mathfrak{N}')$. This contradiction completes the proof.

From theorem 3 and from theorem on isomorphic models it follows at once theorem 1.

Let us notice that theorem 3 and, moreover, theorem 1 remain true when the set S is denumerable. The condition stating that $\overline{\overline{T(X)}} \leqslant \aleph_0$ can be then omitted.

6. Completeness in the Robinson's sense

This concept was introduced by A. Robinson in the monograph [15]. A system X is complete in the Robinson's sense, in symbols, $X \in \boldsymbol{R}$, if and only if either for any model $\mathfrak{M}$ if $X \subset E(\mathfrak{M})$, then $A \in E(\mathfrak{M})$, or for any model $\mathfrak{M}$ if $X \subset E(\mathfrak{M})$, then $A \notin E(\mathfrak{M})$, where A is an arbitrary sentence in S. This concept was also used by A. Church in [3]. The following theorem holds:

THEOREM 4. $\boldsymbol{R} = \mathbf{T}$.

The concept $\boldsymbol{R}$ can be weakened as follows: $X \in \boldsymbol{R}_{\mathfrak{m}}$, if and only if either for any model $\mathfrak{M}$ of power $\mathfrak{m}$ if $X \subset E(\mathfrak{M})$, then $A \in E(\mathfrak{M})$, or for any model $\mathfrak{M}$ of power $\mathfrak{m}$ if $X \subset E(\mathfrak{M})$, then $A \notin E(\mathfrak{M})$, where A is an arbitrary sentence in S. It is easy to observe that $\boldsymbol{R}_{\mathfrak{m}} = \mathbf{T}_{\mathfrak{m}}$.

7. Completeness under provability or completeness in the Gödel's sense

This concept was suggested by K. Gödel in [5] in 1930. A set X is complete in the Gödel's sense, in symbols, $X \in \boldsymbol{G}$, if and only if $Cn(X) = \bigcap \{E(\mathfrak{M}) \colon \mathfrak{M}\}$. It is known from [5] that $Cn(\emptyset) \in \boldsymbol{G}$. Hence the concept $\boldsymbol{G}$ is non-empty. Let us also observe that this concept is disjoined with the concept $\mathfrak{B}$.

The concept $\boldsymbol{G}$ can be weakened as follows: $X \in \boldsymbol{G}_{\mathfrak{m}}$ if and only if $Cn(X) = \bigcap \{E(\mathfrak{M}) \colon \mathfrak{M}, \overline{\overline{\mathfrak{M}}} = \mathfrak{m}\}$. The following theorem holds:

THEOREM 5. *$Cn(\{\delta_n\}) \in \boldsymbol{G}_n$ for any n.*

Proof. Let us assume that $A \notin Cn(\{\delta_n\})$. According to Gödel-Malcev's completeness theorem there exists a model $\mathfrak{N}$ such that $\{\delta_n, \neg A\} \subset E(\mathfrak{N})$. This means that $\overline{\overline{\mathfrak{N}}} = n$ and $A \notin E(\mathfrak{N})$, i.e., $A \notin \bigcap \{E(\mathfrak{M}) \colon \mathfrak{M}, \overline{\overline{\mathfrak{M}}} = n\}$. Therefore $\bigcap \{E(\mathfrak{M}) \colon \mathfrak{M}, \overline{\overline{\mathfrak{M}}} = n\} \subset Cn(\{\delta_n\})$. The converse inclusion is obvious.

The proved theorem testifies that the concept of completeness in the Gödel's sense is non-empty in any finite power. This concept is non-empty also in infinite powers. Namely, the following theorem holds:

THEOREM 6. *$\bigcup_{n=1}^{\infty} Cn(\{\sigma_n\}) \in \boldsymbol{G}_{\mathfrak{m}}$ for any $\mathfrak{m} \geqslant \aleph_0$.*

Proof. Let us assume that $A \notin Cn\big(\bigcup_{n=1}^{\infty} Cn(\{\sigma_n\})\big)$. Let us observe that for any n we have that $\sigma_n \in Cn\big(\bigcup_{n=1}^{\infty} Cn(\{\sigma_n\}) \cup \{\neg A\}\big)$. Let us also observe that the set of the constant terms occurring in the sentences composing the set $Cn(\bigcup_{n=1}^{\infty} Cn(\{\sigma_n\}) \cup \{\neg A\})$ is at most denumerable. According to the theorem cited at the beginning it follows from this that there exists a model $\mathfrak{M}$ of the power $\mathfrak{m}$ such that $A \notin E(\mathfrak{M})$ that is $A \notin \bigcap \{E(\mathfrak{M})\colon \mathfrak{M}, \overline{\overline{\mathfrak{M}}} = \mathfrak{m}\}$. Therefore $\bigcap \{E(\mathfrak{M})\colon \mathfrak{M}, \overline{\overline{\mathfrak{M}}} = \mathfrak{m}\} \subset \subset Cn(\bigcup_{n=1}^{\infty} Cn(\{\sigma_n\}))$. On the other hand, let us assume that $A \in Cn(\bigcup_{n=1}^{\infty} Cn(\{\sigma_n\}))$ and that the model $\mathfrak{M}$ is of any infinite power $\mathfrak{m}$. Hence we get that for any n $\sigma_n \in E(\mathfrak{M})$, that is, $Cn(\bigcup_{n=1}^{\infty} Cn(\{\sigma_n\})) \subset \subset E(\mathfrak{M})$, from which it follows that $A \in \bigcap \{E(\mathfrak{M})\colon \mathfrak{M}, \overline{\overline{\mathfrak{M}}} = \mathfrak{m}\}$. This proves the converse inclusion.

8. Completeness in the Kemeny's sense.

J. Kemeny introduced in [9] in 1948 the concept of completeness which we symbolize by $\boldsymbol{K}$ and define it as follows: $X \in \boldsymbol{K}$, if and only if there exists at least one model $\mathfrak{M}$ such that $X = E(\mathfrak{M})$. It is easy to see that the equality $\boldsymbol{K} = \mathfrak{S} \cap \mathfrak{W} \cap \mathfrak{B}$ holds.

References

[1] *Bemerkung der Redaktion, Supplement to paper of T. Skolem „Über die Nicht-charakterisierbarkeit der Zahlenreihe mittels endlich oder abzählbar unendlich vieler Aussagen mit ausschliesslichen“.*
Fundamenta Mathematicae, 33(1934), 161.

[2] Cantor G.: *Beiträge zur Begründung der transfiniten Mengenlehre*, § 11.
Mathematische Annalen, 46(1895), 510–512.

[3] Church A.: *Introduction to mathematical logic.* Vol. I.
Princeton 1956.

[4] Dedekind R.: *Was sind und was sollen die Zahlen?*
Braunschweig 1888.

[5] Gödel K.: *Die Vollständigkeit der Axiome der logischen Funktionenkalküls.*
Monatshefte für Mathematik und Physik, 37(1930), 349–360.

[6] Grzegorczyk A.: *Zarys logiki matematycznej.*
Wydanie drugie. Warszawa 1969.

[7] Grzegorczyk A.: *On the concept of categoricity.*
Studia Logica, 13(1962), 39–65.

[8] Huntington E.: *A set of postulates for ordinary complex algebra.*
Transactions of the American Mathematical Society, 6(1905), 209–229.

[9] Kemeny J.: *Models of logical systems.*
Journal of Symbolic Logic, 13(1948), 16–30.

[10] ŁOŚ J.: *On the categoricity in power of elementary deductive systems and related problems.* Colloquium Mathematicum, 3(1954), 58–62.

[11] MAL'CEV A.: *Untersuchungen aus dem Gebiete der mathematischen Logik.* Matematičeskij Sbornik, 1(1936), 323–336.

[12] MOSTOWSKI A.: *Logika matematyczna.* Warszawa–Wrocław 1948.

[13] MOSTOWSKI A.: *Współczesny stan badań nad podstawami matematyki.* Prace Matematyczne, I, 1(1955), 13–55.

[14] POST E.: *Introduction to a general theory of elementary propositions.* American Journal of Mathematics, 43(1921), 163–185.

[15] ROBINSON A.: *On the metamathematics of algebra.* Amsterdam 1951.

[16] SURMA S.: *Some metamathematical equivalents of axiom of choice.* Universitas Iagellonica Acta Scientiarum Litterarumque, CLXXIV, Schedae Logicae, 3(1968), 71–80.

[17] TARSKI A.: *Über einige fundamentale Begriffe der Metamathematik.* Comptes Rendus des Séances de la Société des Sciences et des Lettres de Varsovie, Classe III (1930), 22–29.

[18] TARSKI A.: *Fundamentale Begriffe der Methodologie der deduktiven Wissenschaften I.* Monatshefte für Mathematik und Physik, 37(1930), 361–404.

[19] TARSKI A.: *Pojęcie prawdy w językach nauk dedukcyjnych.* Travaux de la Société des Sciences et des Lettres de Varsovie, Classe III, 34(1933), 116.

[20] TARSKI A.: *Z badań metodologicznych nad definiowalnością terminów.* Przegląd Filozoficzny, 37(1934), 438–460.

[21] TARSKI A.: *Grundzüge des Systemenkalküls.* Erster Teil. Fundamenta Mathematicae, 25(1935), 503–526.

[22] TARSKI A.: *Some notions and methods on the borderline of algebra and metamathematics.* Proceedings of the International Congress of Mathematicians, Combridge, Massachusetts, U. S. A., 1950, vol. 1, 1952, 705–720, Providence.

[23] VAUGHT R.: *Applications of the Löwenheim-Skolem-Tarski theorem to problems of completeness and decidability.* Indagationes Mathematicae, 16(1954), 467–472.

[24] VEBLEN O.: *A system of axioms for geometry.* Transactions of the American Mathematical Society, 5(1904), 343–384.

www.ingramcontent.com/pod-product-compliance
Lightning Source LLC
Chambersburg PA
CBHW051116200326
41518CB00016B/2518
* 9 7 8 1 9 3 8 4 2 1 2 6 6 *